AP® CALCULUS AB/BC
ALL ACCESS®

D1540540

Stu Schwartz
Chestnut Hill College
Philadelphia, Pa.

Research & Education Association
Visit our website: www.rea.com

Research & Education Association
61 Ethel Road West
Piscataway, New Jersey 08854
E-mail: info@rea.com

AP® CALCULUS AB/BC ALL ACCESS®

Published 2015
Copyright © 2014 by Research & Education Association, Inc.
All rights reserved. No part of this book may be reproduced in
any form without permission of the publisher.

Printed in the United States of America

Library of Congress Control Number 2013941782

ISBN-13: 978-0-7386-1084-9
ISBN-10: 0-7386-1084-4

Contents

Chapter 1: Welcome to REA's All Access for AP Calculus AB/BC   1

Chapter 2: Strategies for the Exams   7

Chapter 3: A Precalculus Review   27

AP Calculus AB

Chapter 4: Basic Calculus Concepts and Limits — 57

Chapter 5: Derivatives — 81

Chapter 6: Applications of Differentiation — 119

Chapter 7: Integration 155

Chapter 8: Applications of Integration 191

AP Calculus BC

Chapter 9: Additional Limit and Integration Problems 229

Chapter 10: Additional BC Calculus Applications   247

Chapter 11: Sequences and Series   269

AP Calculus AB Practice Exam
(also available online at *www.rea.com/studycenter*)   303

AP Calculus BC Practice Exam
(also available online at *www.rea.com/studycenter*)   353

Answer Sheets   403

Index   405

About Our Author

Stu Schwartz has been teaching mathematics since 1973. For 35 years he taught in the Wissahickon School District, in Ambler, Pennsylvania, specializing in AP Calculus AB and BC and AP Statistics. He currently teaches statistics at Chestnut Hill College in Philadelphia. Mr. Schwartz received his B.S. degree in Mathematics from Temple University, Philadelphia.

Mr. Schwartz was a 2002 recipient of the Presidential Award for Excellence in Mathematics Teaching and also won the 2007 Outstanding Educator of the Year Award for the Wissahickon School District.

Outside the classroom, Mr. Schwartz does consulting work with the Math and Science Partnership of Greater Philadelphia, focusing on factors affecting success in college in STEM (science, technology, engineering, math) degrees. He is well-known in the AP community, having developed the website *www.mastermathmentor.com*, which is geared toward helping educators teach AP Calculus, AP Statistics, and other math courses. Mr. Schwartz is always looking for ways to provide teachers with new and innovative teaching materials, believing that it should be the goal of every math teacher not only to teach students mathematics, but also to find joy and beauty in math as well.

Mr. Schwartz lives with his cat, Newton, in suburban Philadelphia. He loves to bike and play piano in restaurants.

Thanks to my brother, Jerry, who originally kindled my interest in math and always manages to spot small errors in my work. Thanks also to my special friend and business partner, Ted Tyree, for always inspiring me and helping me feel that retiring provided me with an opportunity to share my enthusiasm for learning mathematics with the world.

About REA

Founded in 1959, Research & Education Association (REA) is dedicated to publishing the finest and most effective educational materials—including study guides and test preps—for students in middle school, high school, college, graduate school, and beyond.

Today, REA's wide-ranging catalog is a leading resource for students, teachers, and other professionals. Visit *www.rea.com* to see a complete listing of all our titles.

Acknowledgments

We would like to thank Pam Weston, Publisher, for setting the quality standards for production integrity and managing the publication to completion; John Cording, Vice President, Technology, for coordinating the design and development of the REA Study Center; Larry B. Kling, Vice President, Editorial, for overall direction; Diane Goldschmidt, Managing Editor, for coordinating development of this edition; Transcend Creative Services for typesetting this edition; and Christine Saul, Senior Graphic Designer, for designing our cover.

In addition, we thank Jim Sears of Cape Cod Academy, and Mel Friedman, M.S., for technically reviewing the manuscript.

Welcome to REA's All Access for AP Calculus AB/BC

A new, more effective way to prepare for your AP exam

There are many different ways to prepare for an AP exam. What's best for you depends on how much time you have to study and how comfortable you are with the subject matter. To score your highest, you need a system that can be customized to fit you: your schedule, your learning style, and your current level of knowledge.

This book, and the free online tools that come with it, will help you personalize your AP prep by testing your understanding, pinpointing your weaknesses, and delivering flashcard study materials unique to you.

Let's get started and see how this system works.

How to Use REA's AP All Access

The REA AP All Access system allows you to create a personalized study plan through three simple steps: targeted review of exam content, assessment of your knowledge, and focused study in the topics where you need the most help.

Here's how it works:

Review the Book	Study Chapter 3 to be certain that you are strong on the essential Precalculus topics for AP Calculus. Then, study the topics tested on the AP exam and learn proven strategies that will help you tackle any question you may see on test day. This approach will maximize your score.
Test Yourself & Get Feedback	As you review the book, test yourself. Score reports from your free online tests and quizzes give you a fast way to pinpoint what you really know and what you should spend more time studying.
Improve Your Score	Armed with your score reports, you can personalize your study plan. Review the parts of the book where you are weakest, and use the REA Study Center to create your own unique e-flashcards, adding to the 100 free cards (60 for Calculus AB; 40 for Calculus BC) included with this book.

Finding Your Strengths and Weaknesses: The REA Study Center

The best way to personalize your study plan and truly focus on the topics where you need the most help is to get frequent feedback on what you know and what you don't. At the online REA Study Center, you can access three types of assessment: end-of-chapter quizzes, mini-tests, and a full-length practice test. Each of these tools delivers a detailed score report that follows the topics set by the College Board.

✔ 9 End-of-Chapter Quizzes

Short online quizzes are available throughout the review and are designed to test your immediate grasp of the topics just covered.

✔ 3 Mini-Tests

Two online mini-tests for Calculus AB students (and one online mini-test for Calculus BC students) cover what you've studied in each major section of the book. These tests are like the actual AP exam, only shorter, and will help you evaluate your overall understanding of the subject material.

✔ 2 Full-Length Practice Tests

After you've finished reviewing the book, take our full-length exam for Calculus AB or BC (or both) to practice under test-day conditions. Available both in this book and online at the REA Study Center (*www.rea.com/studycenter*), these tests give you the most complete picture of your strengths and weaknesses. We strongly recommend that you take the online version of the exam for the added benefits of timed testing, automatic scoring, and a detailed score report.

Improving Your Score: e-Flashcards

With your score reports from our online quizzes and practice tests, you'll be able to see exactly which topics you need to review. Use this information to create your own flashcards for the areas where you are weak. And, because you will create these flashcards through the REA Study Center, you'll be able to access them from any computer or smartphone.

Not quite sure what to put on your flashcards? Start with the 100 free cards (60 for Calculus AB; 40 for Calculus BC) included when you buy this book.

After the Full-Length Practice Test: *Crash Course*

After finishing this book and taking our full-length practice exam, pick up REA's *Crash Course for AP Calculus AB & BC*. Use your most recent score reports to identify any areas where you are still weak, and turn to the *Crash Course* for a rapid review presented in a concise outline style.

REA's Suggested 8-Week Study Plan for AP Calculus AB

Depending on how much time you have until test day, you can expand or condense our eight-week study plan as you see fit.

To score your highest, use our suggested study plan and customize it to fit your schedule, targeting the areas where you need the most review.

	Review 1-2 hours	Quiz 20 minutes each	e-Flashcards Anytime, Anywhere	Mini-Test 30 minutes	Full-Length Practice Test 3 hours, 15 minutes
Week 1	Chapters 1, 2				
Week 2	Chapter 3	Quiz 1	Access your e-flashcards from your computer or smartphone whenever you have a few extra minutes to study. Start with the 100 free cards (60 for Calculus AB; 40 for Calculus BC) included when you buy this book. Personalize your prep by creating your own cards for topics where you need extra study.		
Week 3	Chapter 4	Quiz 2			
Week 4	Chapter 5	Quiz 3			
Week 5	Chapter 6	Quiz 4		Mini-Test 1	
Week 6	Chapter 7	Quiz 5			
Week 7	Chapter 8	Quiz 6		Mini-Test 2	
Week 8					Full-Length Practice Exam (Just like test day)

Need even more review? Pick up a copy of REA's *Crash Course for AP Calculus AB/BC*, a rapid review presented in a concise outline style. Get more information about the *Crash Course* series by visiting *www.rea.com*.

REA's Suggested 8-Week Study Plan for AP Calculus BC

Depending on how much time you have until test day, you can expand or condense our eight-week study plan as you see fit. You may be learning some of these topics at the same time you are doing your review. That's okay. The more exposure to these topics that you get, the easier it will be for you to do the problems.

To score your highest, use our suggested study plan and customize it to fit your schedule, targeting the areas where you need the most review.

Most schools offer Calculus AB and Calculus BC as separate courses to be taken in consecutive years. If you are taking Calculus AB and BC in one course, you will need more study time. It is suggested that you consolidate the time spent on the AB topics in order to allow yourself additional time to spend on the BC topics.

	Review 1-2 hours	Quiz 20 minutes each	e-Flashcards Anytime, Anywhere	Mini-Test 30 minutes	Full-Length Practice Test 3 hours, 15 minutes
Week 1	Chapters 1, 2, 3	Quiz 1	Access your e-flashcards from your computer or smartphone whenever you have a few extra minutes to study.		
Week 2	Chapters 4, 5	Quizzes 2, 3			
Week 3	Chapter 6	Quiz 4		Mini-Test 1	
Week 4	Chapters 7, 8	Quizzes 5, 6		Mini-Test 2	
Week 5	Chapter 9	Quiz 7	Start with the 100 free cards (60 for Calculus AB; 40 for Calculus BC) included when you buy this book. Personalize your prep by creating your own cards for topics where you need extra study.		
Week 6	Chapter 10	Quiz 8			
Week 7	Chapter 11	Quiz 9		Mini-Test 3	
Week 8					Full-Length Practice Exam (Just like test day)

Need even more review? Pick up a copy of REA's *Crash Course for AP Calculus AB/BC*, a rapid review presented in a concise outline style. Get more information about the *Crash Course* series by visiting *www.rea.com*.

Test-Day Checklist

✓ Get a good night's sleep. You perform better when you're not tired.

✓ Wake up early.

✓ Dress comfortably. You'll be testing for hours, so wear something casual and layered.

✓ Eat a good breakfast.

✓ Bring these items to the test center:

- Several sharpened No. 2 pencils
- Admission ticket
- Two pieces of ID (one with a recent photo and your signature)
- A noiseless wristwatch to help pace yourself

✓ Arrive at the test center early. You will not be allowed in after the test has begun.

✓ Relax and compose your thoughts before the test begins.

Remember: eating, drinking, smoking, cellphones, dictionaries, textbooks, notebooks, briefcases, and packages are all prohibited in the test center.

Strategies for the Exams

Content of the AP Calculus AB & BC Exams

Although the content of the Calculus AB and BC exams changes slightly from year to year, the following outline shows the topics found on the exams.

I. Functions and Limits

   Limits (graphical and algebraic)
   Continuity
   Parametric functions*
   Polar functions*
   Vector-valued functions*
   L'Hospital's Rule* (expected to be part of the AB curriculum in 2014)

II. Derivatives

   Rates of change
   Derivative definition
   Differentiation techniques for algebraic functions
   Chain rule
   Implicit differentiation
   Derivatives of inverse functions
   Differentiability

III. Applications of Differentiation

   Related rates
   Straight-line motion

Items marked with an asterisk (*) are found *only* on the BC exam. All other topics are common to both exams.

Function analysis

The Intermediate and Mean Value Theorems

Absolute extrema

Optimization

IV. Integrals

Anti-differentiation rules

u-substitution

The definite integral as area

The accumulation function

Riemann sums and the trapezoidal rule

Fundamental theorem of calculus

Integration by parts*

Integration using partial fractions*

Improper integrals*

V. Application of Integration

Area

Average value

Accumulated change

Straight line motion

Volume

Arc length*

Solving differential equations by separation of variables

Growth and decay

Slope fields

Euler's method*

Logistic growth*

VI. Infinite Series

Convergence tests*

Radius/interval of convergence*

Taylor polynomial approximations*

Error bounds*

Taylor and Maclaurin series*

Functions defined as power series*

Items marked with an asterisk (*) are found *only* on the BC exam. All other topics are common to both exams.

Format of the Exams

The Multiple-Choice Section

Each exam begins with multiple-choice questions. Section I, Part A contains 28 multiple-choice questions to be completed in 55 minutes. No calculator is allowed on this part. Section I, Part B contains 17 multiple-choice questions to be completed in 50 minutes. A graphing calculator is required for this part. The problems in Part A are numbered 1–28. However, for unknown reasons, the problems in Part B are numbered 76–92. Each multiple-choice question is followed by five answer choices, (A) through (E).

The Free-Response Section

After time is called on the multiple-choice section, you'll get a short break before diving into the free-response section. The section requires you to produce a total of six written responses, usually with multiple parts, in 90 minutes. Part A contains 2 problems and these require a graphing calculator; this part must be completed in 30 minutes. Part B contains 4 problems and no calculator is allowed. The time allotted for Part B is 1 hour.

The entire exam is summarized in the table below:

Section	Part	Number of Problems	Problem Numbers	Type	Calculator*	Time
I	A	28	1 – 28	Multiple-choice	No	55 min
	B	17	76 – 92	Multiple-choice	Required	50 min
II	A	2	1 – 2	Free response	Required**	30 min
	B	4	3 – 6	Free response	No	60 min
Total						3 hours, 15 minutes

* Just because a calculator is required doesn't mean it will need to be used on all problems.

** If you finish Section II, Part B early, you may return to work on the questions in Section II, Part A. However, you will not be able to use your calculator.

What's the Score?

Although the scoring process for the AP exam may seem quite complex, it boils down to two simple components: your multiple-choice score plus your free-response score. The multiple-choice section accounts for 50% of your overall score and is generated by awarding one point toward your "raw score" for each question you've answered correctly.

The free-response section accounts for the remaining 50% of your total score. Trained graders read students' written responses and assign points according to grading rubrics. Each free-response problem is worth 9 points and the number of points you accrue out of the total points possible will form your score on the free-response section.

The test maker reports AP scores on a scale of 1 to 5. Although individual colleges and universities determine what credit or advanced placement, if any, is awarded to students at each score level, these are the assessments typically associated with each numeric score:

5 Extremely well qualified
4 Well qualified
3 Qualified
2 Possibly qualified
1 No recommendation

Calculus AB Subscore for the Calculus BC Exam

The College Board provides AP Calculus BC test-takers with two scores: the score you received for the Calculus BC exam and the grade you would have received if you had taken the Calculus AB exam. This is called the AB subscore and is based on your performance on the portion of the exam devoted to Calculus AB topics. If you are familiar with the more complex BC topics, you have nothing to lose by taking the BC exam because of the AB subscore.

The Focus of Our Review Chapters

This book contains an AP Calculus AB and BC course review, which is meant to complement your AP Calculus textbook, and while it covers all types of problems you will see in the AP Calculus exams, it is by no means exhaustive. You will, however, be well prepared for the AP exam by using our review along with your textbook.

Your calculus textbook most likely gives a lot of practice problems on individual calculus skills. Many problems give specific instructions: e.g., take the derivative, find the integral, find relative extrema, determine the limit, etc. The review problems in this book concentrate more on problems where calculus skills are combined and where you are not told specifically what skills to use. For example, a word problem that asks about the time when the value of a house is increasing the fastest, uses calculus skills taught by a textbook, but requires students to understand the underlying concepts behind them. In this case, you need to know that the underlying idea is the derivative of the function describing the cost of the house and maximizing it requires you to set it equal to zero. This makes the overall theme of the problem function analysis.

So, while other books may focus on computational skills in calculus, this book was specifically written with an eye towards what typical AP exam questions ask.

Special Focus of Our BC Exam

This book provides two full-length practice exams, one for Calculus AB and one for Calculus BC.

Our AB practice exam is modeled after a typical AB exam, while our BC practice exam specifically emphasizes the BC curriculum. We do this because the BC exam is a rather different animal from any other AP exam. After all, Calc BC is not so much the next course after AB as it is an extension, or continuation, of AB. Because BC embeds all AB topics in addition to its own unique content, you'll actually get an AB subscore when you sit for the BC exam. What's taught in the BC course, and thus what you need to know for the BC exam, equates with what's typically covered in a college Calc II course.

Here's a boiled-down snapshot of what AB vs. BC looks like:

Topic	AB	BC
Continuity & Limits	X	X*
The Derivative	X	X
Applications of the Derivative	X	X
The Integral	X	X*
Applications of the Integral	X	X*
Differential Equations	X	X*
Polar and Parametric Functions		X
Vector-Valued Functions		X
Infinite Series		X

* indicates topic includes both basic and advanced material

What the topic list in itself can't effectively show, however, is the special challenge BC students face in having to tackle more advanced topics in integration, sequences, and function approximation. This, then, sets the stage for them to make the leap to Calc III (Multivariate Calculus or Linear Algebra) as their first college math class. Here's the rub: The BC course will typically go at a faster clip than AB and require more independent out-of-class work. When it comes to the exam, BC asks tougher questions than AB. So in our BC practice exam, we bulk you up with more problems that test BC topics exclusively.

General Strategies for the AP Calculus Exams

The AP Calculus exams are challenging. You have a total of 3 hours and 15 minutes to demonstrate your calculus knowledge and hopefully convince colleges that you have mastered the material so well that you can skip the introductory calculus course. Every concept in the course is tested, some several times. You need to not just perform typical calculus computations, but also be able to describe your thought process using sentences and proper mathematical notation. You need to be able to write so graders can understand what you say. And, you need to be able to work under time pressure.

Here are some general strategies to keep in mind as you prepare for your AP Calculus exam:

Become comfortable with the format of the exam. Stay calm and pace yourself. After simulating the test a few times, you will know what to expect on test day, and you'll relieve your test anxiety.

Work quickly and steadily. On the multiple-choice questions, you have about 2 minutes per problem on the non-calculator section and about 3 minutes per problem on the calculator section. For the free-response problems, you have about 15 minutes per problem. It is important to be aware of your time. Spending 10 minutes on a multiple-choice problem and ultimately getting it correct might mean that you will not complete the section. If you feel that you are spending too much time on a problem, simply guess (eliminating any choices you know are wrong) and move on. If you have time left, come back to the problem.

Be aware of test vocabulary. Words such as *always, every, none, only,* and *never* indicate there should be no exceptions to the answer you choose. Words like *generally, usually, sometimes, seldom, rarely,* and *often* indicate that there may be exceptions to your answers.

Learn the directions and format for each section of the test. Familiarizing yourself with the directions and format of the exam will save you valuable time on test day.

Wear a watch. Rather than being dependent on having a test administrator call out times, you can keep track of your time yourself. Once the sections begin, write down when, according to your watch, the exam will end. Also remember that the use of electronic devices, like cell phones, is not allowed in the exam room. So you won't be able to check the time on your cell phone if that is your usual practice.

Section I: Strategies for the Multiple-Choice Section

The AP exam is a standardized test, so each version of the test from year to year must share many similarities in order to be fair. That means that you can always expect certain things to be true about your AP Calculus exam.

Which of the following phrases accurately describes a multiple-choice question on the AP Calculus exam?

(A) always has 5 choices

(B) may rely on a graph, table, or other visual stimulus

(C) may ask you to determine an incorrect conclusion from given information

(D) may test calculus theory rather than **computational basics**

(E) all of the above *

Did you pick "all of the above?" Good job!

What does this mean for your study plan? You should focus more on application and interpretation of limits, derivatives, and integrals than on the nuts and bolts of actually finding them. Multiple-choice questions usually have simple functions to differentiate. But rather than focusing on the process of differentiation, the challenge will be when you are asked what the derivative represents in the context of a problem situation. In problems like these, there are few or no computations. These interpretation problems show up typically on problems that ask you to use Roman numerals to organize ideas into categories. Not sure what this type of question might look like? Let's examine a typical Roman numeral item:

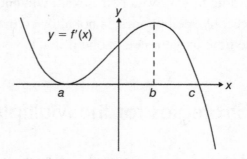

The graph above is $y = f'(x)$, the derivative of the function f. Which of the following statements is correct?

---

* Of course, on the actual AP Calculus exam, you won't see any choices featuring "all of the above" or "none of the above."

I. f is decreasing on (b, c)

II. f has a relative minimum at $x = a$

III. f has a relative maximum at $x = c$

(A) I only

(B) II only

(C) III only

(D) I and II

(E) no statements are correct

First, it is important to look at the given information. You are given a graph of $y = f'(x)$, not $f(x)$. This gives you information about the derivative of the function f. Interpreting the meaning of the derivative allows you to answer the question. Knowing that when the derivative is positive, the function is increasing, eliminates choice (A) (and also choice (D)). Knowing that a relative minimum occurs when $f'(x)$ switches from negative to positive eliminates choice (B). Knowing that a relative maximum occurs when $f'(x)$ switches from positive to negative tells you that choice (C) must be the answer.

Students typically choose choice (D) in a problem like this because they assume that they are looking at the function and not the derivative of the function. Be sure to carefully read the problem and focus on what you are given.

Here are some strategies for the multiple-choice section:

Be aware of time. Allow about 2 minutes per question on Section A (non-calculator) and 3 minutes per question on section B (calculator). If you see that your solution is getting too involved, you should rethink your approach as none of the AP questions requires complicated solutions.

If you skip a problem, be sure to skip the corresponding problem number on your answer sheet. Always be sure that the problem you are answering matches the problem number of the bubble sheet.

Develop a simple set of codes that tell you what to do if you are not sure of how to solve a problem. For instance, you might want to circle the question number if you feel that more time would help you answer the question. If you finish the multiple-choice section early, you can return to the circled questions and give them the time needed. Consider putting an X through a problem you should not spend any more time on.

Be sure you answer every problem. Don't leave any problem blank, even if you have to guess randomly. You might just guess the right answer. Remember, there is no guessing penalty. You earn points for every question you answer correctly and no points are deducted for incorrect answers.

Get right what you are sure you know. If you see a problem and you absolutely know how to solve it, you must get it correct. Every correct problem is gold!

Read all the possible answers. Just because you think you have found the correct response, do not automatically assume that it is the best answer. Read through each answer choice to be sure that you are not making a mistake by jumping to conclusions. Let's look at a fairly simple algebraic question as an example:

Solve the equation $x^2 - 1 = 3$.

(A) 2
(B) ±2
(C) $\sqrt{2}$
(D) ±$\sqrt{2}$
(E) 4

Answers (C), (D), and (E) are distractors—answers that careless mistakes will lead to. Answer (A) is either a conceptual or careless error. The correct answer is, of course, (B).

Use the process of elimination. Go through each answer to a problem and eliminate as many of the answer choices as possible by putting a line through them. Developing incorrect answers (also known as distractors) is difficult for the test creators and there are usually one or more answers that logically are impossible. For instance, if you were asked to find the minimum value of $x^2 + 4$, it would be impossible for the answer to be a negative number.

Remember that every answer is independent. Just because you have four C's in a row doesn't mean that the next answer cannot be (C). Just because you haven't had the answer (E) for a long time doesn't mean that it is time for an answer to be (E).

Be neat. Many students take the philosophy that since their work won't be seen, they can be sloppy. Sloppy work, though, can lead to careless mistakes and the test creators know exactly the types of careless mistakes that students typically make. Also, if you come back to the problem later in the exam, you want to be in a position to look at work that is easy to understand.

Don't overthink a problem. More often than not, your first intuition will be correct.

Answer the question. If you were asked to find the area of a square with side x and your work yielded $x = 3$, be sure to choose the answer 9, and not 3. The question asked you to find the area of the square (x^2), not x. Underlining both the important terms in the problems as well as the actual question being asked will eliminate the possibility of errors like this.

Analyzing a Multiple-Choice Question

A pyramid with a square base, shown below, has a volume given by $V = \frac{1}{3}s^2h$ where s is the length of the side of the square and h is the base. At the time when the area of the base is 16 cm², its area is increasing at the rate of $\frac{24 \text{ cm}^2}{\text{min}}$ and the height of the pyramid is 6 cm, decreasing at the rate of $\frac{3 \text{ cm}}{\text{min}}$. How fast is the volume of the pyramid changing at this time?

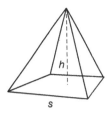

(A) decreasing at $\dfrac{40 \text{ cm}^3}{\text{min}}$

(B) increasing at $\dfrac{64 \text{ cm}^3}{3 \text{ min}}$

(C) decreasing at $\dfrac{16 \text{ cm}^3}{\text{min}}$

(D) increasing at $\dfrac{64 \text{ cm}^3}{\text{min}}$

(E) increasing at $\dfrac{96 \text{ cm}^3}{\text{min}}$

SOLUTION: This is clearly a related-rates problem as you are given a geometric figure with certain dimensions changing. Problems like this require organization and keeping track of all variables. In this problem, our goal is to find the change of volume: $\dfrac{dV}{dt}$.

$$V = \frac{1}{3}s^2h \text{ so } \frac{dV}{dt} = \frac{1}{3}\left(s^2\frac{dh}{dt} + 2sh\frac{ds}{dt}\right)$$

Information given : $A = 16$, $\dfrac{dA}{dt} = 40$, $h = 6$, $\dfrac{dh}{dt} = -3$

Since $A = s^2 = 16$, $s = 4$

and $\dfrac{dA}{dt} = 2s\dfrac{ds}{dt}$ so $40 = 2(4)\dfrac{ds}{dt} \Rightarrow \dfrac{ds}{dt} = 5$

So we have all the necessary information:

$$\frac{dV}{dt} = \frac{1}{3}\left(s^2\frac{dh}{dt} + 2sh\frac{ds}{dt}\right) = \frac{1}{3}\left[4^2(-3) + 2(4)(6)(5)\right]$$

$$\frac{dV}{dt} = \frac{1}{3}(-48 + 240) = 64$$

The volume is increasing at the rate of $\dfrac{64 \text{ cm}^3}{\text{min}}$

The correct answer is (D).

Let's examine the incorrect answer choices individually:

- Choice (A) might have been chosen by a student who could not do implicit differentiation and thought $\dfrac{dV}{dt} = \dfrac{1}{3}(2s)\dfrac{dh}{dt} \cdot \dfrac{ds}{dt}$.

- Choice (B) might have been attractive to a student who forgot the chain rule and thought $\dfrac{dV}{dt} = \dfrac{1}{3}(s^2 + 2sh)$.

- Choice (C) might have been chosen by a student who forgot the product rule partially and thought $\dfrac{dV}{dt} = \dfrac{1}{3}\left(s^2\dfrac{dh}{dt}\right)$

- A student who did not remember that $\dfrac{dt}{dt}$ is negative would likely have chosen (E).

You should also realize that simply by looking at the formula, the area of the base is more important than the height in determining the volume. Since the area of the base is increasing, and the height isn't decreasing that quickly, it is likely that the volume is increasing and, thus, answer choices (A) and (C) can be readily eliminated.

Again, remember to read the question carefully, read ALL of the answer choices, and examine any visual cues that are given. You'll see the results in your AP score!

Section II: Strategies for the Free-Response Section

Free-response questions often provide you with one or more visual stimuli, such as tables, graphs, and figures. You are then asked a series of 3 or 4 questions. The questions usually increase in difficulty and typically answers from one question are used in a subsequent question.

In the Calculus BC exam, usually three of the 6 free-response questions will focus on BC concepts and the other three questions will be the same questions that are in the Calculus AB exam.

Students can be asked to justify an answer. This means that a specific reason using calculus concepts is required to explain the numerical answer. Justifications should be written in sentence form.

Free-response questions often have general themes. Some examples of these themes include function analysis, particle motion, related rates, area and volume, continuity and differentiability, solving differential equations, accumulated change, linear approximation, and infinite series. Within the sub-questions of these problems, a lot of the nuts and bolts of calculus computations and analysis are tested.

Here are some general tips for the free-response section of the AP Calculus exam.

Be aware of time. Allow about 15 minutes per problem on both Part A (calculator) and Part B (no-calculator). While each problem might have multiple parts, no one part should take an inordinate amount of time. If you see that your solution is getting too involved, you should rethink your approach as there will be no problems that necessitate complicated solutions.

Read the entire problem first before beginning to answer any of the parts. You can choose to answer the parts in any order you wish. In this way, you maximize your available time to answer all the parts of a particular problem.

Answer the question being asked. Underline the question and be sure that you answer the question that is asked.

Be neat. This suggestion is doubly important in the free-response section because people are going to read your work. Make it as easy as possible for them. An AP reader will not spend time guessing the intent of your answer. Be sure your work flows and if you are moving from one side of the booklet to another, use an arrow to direct the eyes of the reader.

Don't leave parts of the problem blank. You can usually do something, even if it is writing down the given information in variable form. For instance, if you are told that x is increasing at the rate of 2 ft/sec, and you write that $\frac{dx}{dt} = 2$ without doing anything else, you might earn one point. And, as you know, every point helps.

Match your variables to your diagrams. If there is a diagram with a right triangle and you label the sides a, b, and c, be sure that your work doesn't switch to talking about x, y, and z.

Extend graphs to the edges of the provided axis. Typically, students are asked to graph an approximate solution to a differential equation onto a slope field. You will lose points unless the graphs are extended to the edges showing that the graph continues forever.

Do not waste time erasing wrong solutions. If you change your mind, simply cross out the wrong solution after you have written the correct one. Crossed-out work will not be graded. If you have no better solution, leave the old one there. It might be worth a point or two.

Don't simplify answers. Over the years, you have probably been taught that certain types of simplifications are necessary. You were taught not to leave a radical in the denominator—you had to change $\frac{1}{\sqrt{2}}$ to $\frac{\sqrt{2}}{2}$. That is *not* true on the AP Calculus exams. It is recommended that you not spend time doing these simplifications. For instance, if your answer to finding the equation of a line was $y - 4 = \dfrac{-2}{3}(x - 3)$, don't spend the time changing it to $y = \dfrac{-2}{3}x + 6$. It is not necessary and, if you make a mistake, you will lose a point.

Show all work. Remember that the AP reader is not interested in finding out the answer to the problem. The grader is interested in seeing if you know how to solve the problem. So show all of your work.

Be aware of the words "Justify your answer," "Explain why," and "Show the reasoning that leads to your conclusion." They want you to write a sentence or two that shows your thought process. Write in complete sentences and use the variables that you have established in your solutions. If you are asked to justify your answer, one of the points in the rubric will be for this justification. If you are not asked to justify, don't spend time doing so.

Watch your definite articles. If you were asked to take the first and second derivative of a function and then justify why a function is increasing and you stated "because it is positive," you would lose points because it isn't clear what "it" refers to. Is "it" the function? the derivative? the 2nd derivative? Be sure your statement is clear.

Do not let the points at the beginning keep you from getting the points at the end. For the free-response questions, if you know you can do part (b) without doing part (a), do it. If you need to use an answer from part (a) to answer part (b), make a credible attempt at part (a). Even if your part (a) answer is incorrect, you might still earn points on part (b).

Watch for solutions that are too long. Most problems on the free-response section can be completed in about 7 steps. The testers know that your time is limited. If you find that a solution is getting too convoluted, it may be best to stop (don't erase it) and think of a better way to proceed. If you do, perform the steps and then cross out your previous work. You don't want to waste too much time on something that might be worth 3 points at the cost of not completing the exam.

Watch your use of the equal and the approximation sign. Typical problems ask students to use a tangent line to approximate the value of a function at a point or approximate the value of a definite integral with Riemann sums. Be sure that you use an approximation sign ($\approx$) when asked for an approximation instead of an equal sign or you will lose a point.

Analyzing a Free-Response Question

A particle initially at rest with no initial acceleration moves along the x-axis with $a'(t) = 6 + 4 \cos 2t$. Its position at $t = 0$ is $x = 4$.

a) Determine whether the particle is speeding up or slowing down at $t = \dfrac{\pi}{4}$. Show the reasoning that leads to your conclusion.

b) Write an expression that describes how far the particle traveled between $t = 0$ and $t = \dfrac{\pi}{4}$. Do not evaluate.

SOLUTION: This is clearly a straight-line motion problem. Your given information tells you a', so you need to work backwards to a and then v. You need to perform integration to do part a of this problem. Part b requires you to know the formula for distance traveled. So you need to use $v(t)$ found from part a to answer this question.

a) $a(t) = \displaystyle\int (-6 + 4 \cos 2t)\, dt = -6t + 2 \sin 2t + C_1$

$a(0) = 0 + 2 \sin 0 + C_1 = 0 \Rightarrow C_1 = 0$

$a(t) = -6t + 2 \sin 2t$

$v(t) = \displaystyle\int (-6t + 2 \sin 2t)\, dt = -3t^2 - \cos 2t + C_2$

$v(0) = 0 - 1 + C_2 = 0 \Rightarrow C = 1$

$v(t) = -3t^2 - \cos 2t + 1$

$v\left(\dfrac{\pi}{4}\right) = -3\left(\dfrac{\pi^2}{16}\right) - \cos\dfrac{\pi}{2} + 1 = \dfrac{-3\pi^2}{16} + 1 < 0$

$a\left(\dfrac{\pi}{4}\right) = -6\left(\dfrac{\pi}{4}\right) + 2\sin\dfrac{\pi}{2} = \dfrac{-3\pi}{2} + 2 < 0$

Since $v\left(\dfrac{\pi}{4}\right)$ and $a\left(\dfrac{\pi}{4}\right)$ are both negative, the particle is speeding up.

b) Distance $= \displaystyle\int_0^{\pi/4} |v(t)|\, dt = \int_0^{\pi/4} \left| -3t^2 - \cos 2t + 1 \right| dt$

Let's examine the response:

- The fact that you are asked to show your reasoning means that you have to find expressions for both a and v, evaluate them at $t = \dfrac{\pi}{4}$, and interpret their meaning.

- It is not necessary to actually compute $v\left(\dfrac{\pi}{4}\right)$ and $a\left(\dfrac{\pi}{4}\right)$. All that is needed is their signs.

- Once the signs of $v\left(\dfrac{\pi}{4}\right)$ and $a\left(\dfrac{\pi}{4}\right)$ have been found, the question must be answered and an explanation given.

- If you evaluate these expressions incorrectly, leading you to an incorrect conclusion, you could be awarded points for the justification if your reasoning were sound, based on the answers you had.

- Observe that the answer from part a was used in part b. If the answer from part a were incorrect, but the form of answer b was $\displaystyle\int_{0}^{\pi/4} |v(t)|\, dt$, you could still receive credit for part b.

- Realize that the original position of $x(0) = 4$ was not used in the solution for part b.

- You are not asked to evaluate the expression in part b, so don't spend time doing so.

Calculators

In the construction of the AP Calculus exams, the AP Calculus Exam Development Committee is aware of the wide range of capabilities of graphing calculators. To this end, the committee develops exams that ensure that all students have sufficient calculator capability for all exam problems. Thus, the committee has created some restrictions on calculators that may be used on the AP Calculus exams, which is summarized in the chart below. Test administrators are required to check calculators before the exam.

1.	Calculators allowed	Check the College Board website for the most up-to-date list of sanctioned calculators.
	Calculator features that are not allowed	• QWERTY keyboard • computers • pen-input driven devices • electronic pads • Internet access • non-graphing calculators • apps on cellphones • any device with a camera

(Continued)

(*Continued*)

2.	Memory	Memories will NOT be cleared
3.	Capabilities required	• Plot a function in a viewing window • Find the roots of functions or intersection points of functions • Numerically calculate the derivative of a function • Numerically calculate the value of a definite integral
4.	What work do I show?	The set-up of the problem e.g., Volume $= \pi \int_{0}^{\frac{\pi}{4}} \left(\left(x^2 \right)^2 - \left(\sqrt{\sin x} \right)^2 \right) dx$ Do NOT write the math in calculator syntax: Writing the above as $\pi \text{FNINT} \left(\left(\sqrt{} \left(\sin x \right) \right) \right) \wedge 2 - (x \wedge 2) \wedge 2, x, 0, \frac{\pi}{4}$ would receive no credit. If a free-response problem specifically requires that you show the work for your anti-differentiation on a section where the calculator is permitted, then the calculator can be used only for verifying your algebraic answer.

Just because the calculator *can* be used, does not mean it *must* be used.

Using the Calculator

There are only 4 features of the calculator that you can use without justifications:

1. graphing functions in a particular window;
2. finding zeros (roots) of equations or finding intersections points of two functions;
3. computing the value of derivatives at some value of x; and
4. computing definite integrals.

Other than basic numeric calculations, you may not use any of the calculator's built-in features without mathematical justification. For instance, many calculators have built-in features to find a minimum. While you can use the calculator to explore the behavior of the function, the calculus justification must stand on its own. For that reason, it is worthless to spend time entering programs into your calculator because they cannot be used.

Here are some tips for calculator use:

Have an extra set of batteries on hand. Have them on your desk or table and not in a bag or backpack.

Set your calculator accurate to 3 decimal places and radians. The only time you may want to be in degree mode is if you are taking a trigonometric function of an angle given in decimal degrees.

Do not round partial answers. Store them in your calculator so that you can use them unrounded in further calculations. For instance, if an answer was $x = \dfrac{13}{7}$ and you wrote 1.857 and later you needed to multiply your result by 70, the correct answer would be 130 instead of $70 \cdot 1.857 = 129.990$.

If you use your calculator to solve an equation, write the equation first. An answer without an equation might not get full credit, even if it is correct.

If you are asked to write and find the value of a definite integral, write the integral first. An answer without an integral will not get full credit, even if it is correct.

Do not use "calculator language" in your solutions. Be sure to write mathematics. If you want to find the area under the curve $y = x^2$ from $x = 0$ to $x = 2$, you write $\int_{0}^{2} x^2 dx$. Statements like FNINT(x^2, x, 0, 3) are meaningless to the grader. And don't explain your calculator steps. Statements like "I graphed the functions in a Zoom 6 window and then pressed 2nd CALC 2 to find the root at $x = 3$ and then used the statement nDeriv(Y1, x, 3) to find the derivative" are not math.

Analyzing a Calculator Problem

If f is a function such that $f'(x) = e^{\sin x + \cos x}$, how many critical points does f have on $[0, 4\pi]$?

(A) 0
(B) 1
(C) 2
(D) 3
(E) 4

 **SOLUTION:** This is a function analysis problem. You are given information about the derivative of a function and asked about the function itself. It is similar in nature to the Roman Numeral problem described in the multiple-choice section of this chapter, but focuses on using the calculator.

This is a problem that could appear on either the calculator-active section or the non-calculator section. The calculator approach is to graph $f'(x)$, which is below. The definition of a critical point is where $f'(x) = 0$ or $f'(x)$ does not exist. $f'(x)$ exists everywhere and it should be obvious that $f'(x)$ never touches or crosses the x-axis. Thus, the answer is (A). Choice (E) is a distractor for students who look at the graph and do not realize that they are looking at $f'(x)$ and not $f(x)$.

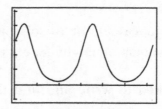

If this problem were part of a free-response question and students were asked to justify their answer, the graph could be part of the justification because plotting a graph in a window is one of the four calculator functions that is allowable on the AP exam. The answer that would receive full credit is: "As shown by the graph of $f'(x)$, it never touches the x-axis. So $f'(x)$ never equals zero and thus $f(x)$ has no critical points."

In addition, realize that this problem does not need a calculator. Since e raised to any power is always positive, then $f'(x)$ must always be positive and never equal zero. Having the use of the calculator can actually make students careless.

A Precalculus Review

You are taking an Advanced Placement Calculus course. It is either Calculus AB or Calculus BC. Let's understand what these unusual names mean. The AP Calculus program started in 1956. There was only one calculus exam given in the early years and it was called "Math." However, once the Calculus program got fully rolling, the courses were split into AB and BC. The first year there were specific AB and BC exams was in 1969. There were three general topics into which all math problems fell:

A Topics: Precalculus concepts. They use no calculus, but are considered necessary for the student to know before entering the study of calculus.

B Topics: Calculus concepts taught in a first-year college calculus course. (In this book, the B topics are found in chapters 4 through 8).

C Topics: Calculus concepts taught in a second-year college calculus course. (In this book, the C topics are found in chapters 9 through 11).

In a typical Calculus AB course, students will see problems including A topics and B topics and, in a BC course, students will see problems including B topics and C topics. Before 2000, there were problems on the AB exam that were strictly A topics—no calculus was required. That is no longer true. In reality, the 45 multiple-choice questions and the 6 free-response questions on the AB exam are B topic questions. They test calculus. Although A topics are not specifically tested, students need to understand them. Just like spelling, while students are not tested specifically on their spelling abilities by the time they get to high school, it is assumed that they know how to spell.

As a review, there are 20 precalculus (A topics) in this chapter that you need to know and master before you start your prep for either AP Calculus exam. This is not meant to be a comprehensive review and if some of these topics are still a mystery to you, ask your teacher for an algebra, trigonometry, or precalculus book to sharpen your skills. The topics reviewed here are not the only ones essential to mastering precalculus, but were chosen because they will more than likely appear in calculus problems.

1. Functions

The lifeblood of precalculus is functions. A function is a set of points (x, y) such that for every x, there is one and only one y. In short, in a function, the x-values cannot repeat and the y-values can repeat.

The notation for functions is either "$y =$" or "$f(x) =$". In the $f(x)$ notation, we are stating a rule to find y given a value of x.

EXAMPLE 1: If $f(x) = x^2 - 2x + 4$, find

a) $f(-5)$;

b) $f(x + h) - f(x)$

SOLUTIONS: a) $f(-5) = (-5)^2 - 2(-5) + 4 = 25 + 10 + 4 = 39$

b) $f(x + h) - f(x) = (x + h)^2 - 2(x + h) + 4 - (x^2 - 2x + 4)$

$$= x^2 + 2xh + h^2 - 2x - 2h + 4 - x^2 + 2x - 4$$

$$= h^2 + 2xh - 2h$$

EXAMPLE 2: If $f(x) = x^2 - 5$ and $g(x) = 2x + 1$, show that $f(g(x)) \neq g(f(x))$.

SOLUTION: $f(g(x)) = f(2x + 1) = (2x + 1)^2 - 5 = 4x^2 + 4x - 4$

$g(f(x)) = g(x^2 - 5) = 2(x^2 - 5) + 1 = 2x^2 - 9$

DIDYOU**KNOW?**

> The term "function" was first coined as a mathematical term in 1673 by mathematician Gottfried Leibniz. In 1734, Leonhard Euler (whose name you will see repeatedly in this book) introduced the familiar $f(x)$ notation. The definition of a function has changed many times since then.

2. Domain and Range

Since questions in calculus usually ask about behavior of functions in intervals, understand that intervals can be written with a description in terms of $<, \leq, >, \geq$ or by using interval notation.

Description	Interval notation
$x > a$	(a, ∞)
$x \geq a$	$[a, \infty)$
$x < a$	$(-\infty, a)$
$x \leq a$	$(-\infty, a]$
$a < x < b$	(a, b) - the open interval
$a \leq x \leq b$	$[a, b]$ - the closed interval
$a \leq x < b$	$[a, b)$
$a < x \leq b$	$(a, b]$
All real numbers	$(-\infty, \infty)$

If a solution is in one interval or the other, interval notation will use the connector $\cup$. So $x \leq 2$ or $x > 6$ would be written $(-\infty, 2] \cup (6, \infty)$ in interval notation.

The domain of a function is the set of allowable x-values. The domain of a function f is $(-\infty, \infty)$ except for values of x which create a zero in the denominator, an even root of a negative number, or a logarithm of a non-positive number.

The range of a function is the set of allowable y-values. Finding the range of functions algebraically isn't as easy (it really is a calculus problem), but visually, it is the interval [lowest possible y-value, highest possible y-value].

EXAMPLE 3: Find the domain of $y = \dfrac{x^2 + 4x + 6}{\sqrt{2x+4}}$.

SOLUTION: $2x + 4 > 0 \Rightarrow x > -2$

EXAMPLE 4: The figure below is the graph of $y = f(x)$. What is the domain and range of the function?

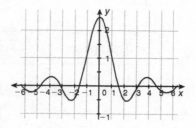

SOLUTION: Domain: $(-\infty, \infty)$
Range: $[-0.5, 2.5]$

3. The Function Toolbox

There are certain graphs that occur all the time in calculus and students should know the general shape of them, where they intersect the x-axis (zeros) and y-axis (y-intercept). We call these graphs the function toolbox.

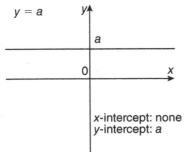

$y = a$

x-intercept: none
y-intercept: a

Domain: $(-\infty, \infty)$
Range: $[a, a]$

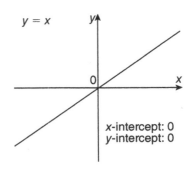

$y = x$

x-intercept: 0
y-intercept: 0

Domain: $(-\infty, \infty)$
Range: $(-\infty, \infty)$

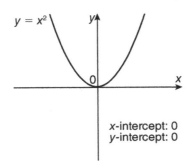

$y = x^2$

x-intercept: 0
y-intercept: 0

Domain: $(-\infty, \infty)$
Range: $[0, \infty)$

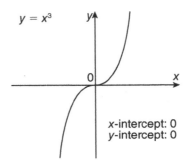

$y = x^3$

x-intercept: 0
y-intercept: 0

Domain: $(-\infty, \infty)$
Range: $(-\infty, \infty)$

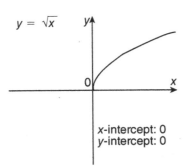

$y = \sqrt{x}$

x-intercept: 0
y-intercept: 0

Domain: $[0, \infty)$
Range: $[0, \infty)$

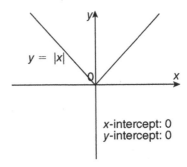

$y = |x|$

x-intercept: 0
y-intercept: 0

Domain: $(-\infty, \infty)$
Range: $[0, \infty)$

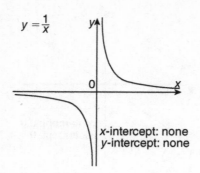

$y = \dfrac{1}{x}$

x-intercept: none
y-intercept: none

Domain: $x \neq 0$
Range: $y \neq 0$

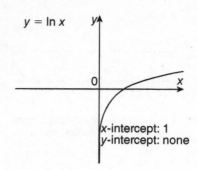

$y = \ln x$

x-intercept: 1
y-intercept: none

Domain: $(0, \infty)$
Range: $(-\infty, \infty)$

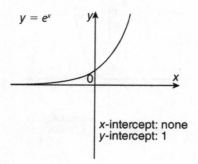

$y = e^x$

x-intercept: none
y-intercept: 1

Domain: $(-\infty, \infty)$
Range: $(0, \infty)$

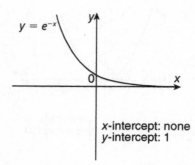

$y = e^{-x}$

x-intercept: none
y-intercept: 1

Domain: $(-\infty, \infty)$
Range: $(0, \infty)$

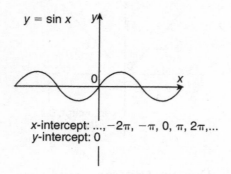

$y = \sin x$

x-intercept: ..., -2π, $-\pi$, 0, π, 2π,...
y-intercept: 0

Domain: $(-\infty, \infty)$
Range: $[-1, 1]$

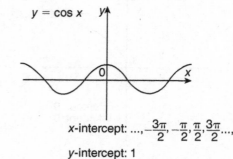

$y = \cos x$

x-intercept: ..., $-\dfrac{3\pi}{2}$, $-\dfrac{\pi}{2}$, $\dfrac{\pi}{2}$, $\dfrac{3\pi}{2}$...,

y-intercept: 1

Domain: $(-\infty, \infty)$
Range: $[-1, 1]$

4. Even and Odd Functions

Functions that are even have the characteristic that for all $a, f(-a) = f(a)$. Even functions are symmetric to the y-axis.

Functions that are odd have the characteristic that for all $a, f(-a) = -f(a)$. Odd functions are symmetric to the origin.

EXAMPLE 5: Of the toolbox functions, which are even, which are odd, and which are neither?

SOLUTION: Even: $y = a, y = x^2, y = |x|, y = \cos x$

Odd: $y = x, y = x^3, y = \dfrac{1}{x}, y = \sin x$

Neither: $y = \sqrt{x}, y = \ln x, y = e^x, y = e^{-x}$

EXAMPLE 6: Determine if $f(x) = x^3 - x^2 + x - 1$ is even, odd, or neither. Justify your answer.

SOLUTION: $f(-x) = -x^3 - x^2 - x - 1 \neq f(x)$, so f is not even.

$-f(x) = -x^3 + x^2 - x + 1 \neq f(-x)$, so f is not odd.

5. Transformation of Graphs

A curve in the form $y = f(x)$, which is one of the basic toolbox functions, can be transformed in a variety of ways. The shape of the resulting curve stays the same, but zeros and y-intercepts might change and the graph could be reversed. The table below describes transformations to a general toolbox function $y = f(x)$ with the parabolic function $f(x) = x^2$ as an example.

Notation	How $f(x)$ changes	Example with $f(x) = x^2$
$f(x) + a$	Moves graph up a units	
$f(x) - a$	Moves graph down a units	
$f(x + a)$	Moves graph a units left	
$f(x - a)$	Moves graph a units right	
$a \cdot f(x)$	$a > 1$: Vertical Stretch	
$a \cdot f(x)$	$0 < a < 1$: Vertical shrink	
$f(ax)$	$a > 1$: Horizontal compress (same effect as vertical stretch)	

(*Continued*)

(Continued)

Notation	How $f(x)$ changes	Example with $f(x) = x^2$
$f(ax)$	$0 < a < 1$: Horizontal elongated (same effect as vertical shrink)	
$-f(x)$	Reflection across x-axis	
$f(-x)$	Reflection across y-axis	

6. Special Factorization

While factoring skills were more important in the days when A topics were specifically tested, students still must know how to factor. The special forms that occur most regularly are:

Difference of squares: $x^2 - y^2 = (x + y)(x - y)$
Perfect squares: $x^2 + 2xy + y^2 = (x + y)^2$
Perfect squares: $x^2 - 2xy + y^2 = (x - y)^2$
Sum of cubes: $x^3 + y^3 = (x + y)(x^2 - xy + y^2)$
Difference of cubes: $x^3 - y^3 = (x - y)(x^2 + xy + y^2)$

EXAMPLE 7: Factor the expression: $36x^2 - 64$.

SOLUTION: Either pull out a common factor first and then use the difference of squares: $36x^2 - 64 = 4(9x^2 - 16) = 4(3x + 4)(3x - 4)$
or use the difference of squares first and then pull out the constant factors:

$36x^2 - 64 = (6x + 8)(6x - 8) = 2(3x + 4)(2)(3x - 4) = 4(3x + 4)(3x - 4)$

EXAMPLE 8: Factor $x^6 - 1$.

SOLUTION: $x^6 - 1 = (x^3 + 1)(x^3 - 1) = (x + 1)(x^2 - x + 1)(x - 1)(x^2 + x + 1)$

7. Linear Equations

Probably the most important concept that is required for differential calculus is that of linear equations. The formulas you need to know completely are:

Slope: Given two points (x_1, y_1) and (x_2, y_2), the slope of the line passing through the points can be written as:

$$m = \frac{\text{rise}}{\text{run}} = \frac{\Delta y}{\Delta x} = \frac{y_2 - y_1}{x_2 - x_1}$$

Slope intercept form: the equation of a line with slope m and y-intercept b is given by $y = mx + b$.

Point-slope form: the equation of a line passing through the point (x_1, y_1) and slope m is given by $y - y_1 = m(x - x_1)$. While you might have preferred the simplicity of the $y = mx + b$ form in your algebra course, the $y - y_1 = m(x - x_1)$ form is far more useful in calculus.

Two distinct lines are parallel if they have the same slope: $m_1 = m_2$.

Two lines are normal (perpendicular) if their slopes are negative reciprocals: $m_1 \times m_2 = -1$.

EXAMPLE 9: Find the equation of the line passing through $(-2, 5)$ with slope $\frac{-1}{2}$.

SOLUTION: $y - 5 = -\frac{1}{2}(x + 2)$ or $y = -\frac{1}{2}x + 4$ or $x + 2y - 8 = 0$.

Note: While your algebra teacher might have required your changing the equation to general form $x + 2y - 8 = 0$, you will find that on the AP test, it is sufficient to leave equations for lines in slope-intercept form $y - 5 = -\frac{1}{2}(x + 2)$ and it is recommended not to waste time changing it.

EXAMPLE 10: Line L passes through $\left(\dfrac{2}{3}, \dfrac{-3}{2}\right)$ and $\left(4, \dfrac{5}{6}\right)$. Find the equation of

the line normal to line L passing through $\left(4, \dfrac{5}{6}\right)$.

SOLUTION: $m_L = \left(\dfrac{\dfrac{5}{6} + \dfrac{3}{2}}{4 - \dfrac{2}{3}}\right)\left(\dfrac{6}{6}\right) = \dfrac{5+9}{24-4} = \dfrac{14}{20} = \dfrac{7}{10}$ so $m_L \perp = \dfrac{-10}{7}$

Normal line: $y - \dfrac{5}{6} = \dfrac{-10}{7}(x-4)$

Working with complex fractions is important and is covered in section 12 in this chapter.

8. Solving Quadratic Equations

Solving quadratics in the form of $ax^2 + bx + c = 0$ usually shows up on the AP exam in the form of expressions that can easily be factored. When you have a quadratic equation, factor it, set each factor equal to zero and solve. If the quadratic equation doesn't factor or if factoring is too time-consuming, use the quadratic formula: $x = \dfrac{-b \pm \sqrt{b^2 - 4ac}}{2a}$. The discriminant $b^2 - 4ac$ will tell you how many solutions the quadratic has:

$$b^2 - 4ac \begin{cases} > 0, \ 2 \text{ real solutions} \\ = 0, \ 1 \text{ real solution} \\ < 0, \ 0 \text{ real solutions (2 imaginary solutions)} \end{cases}$$

EXAMPLE 11: Find the domain of $y = \dfrac{2x-1}{6x^2 - 5x - 6}$.

SOLUTION: $\dfrac{2x-1}{6x^2 - 5x - 6} = \dfrac{2x-1}{(2x-3)(3x+2)}$. The domain is all allowable values of x, which means $(2x - 3)(3x + 2) \neq 0$. Then $2x - 3 \neq 0$ and $3x + 2 \neq 0$. So $x \neq \dfrac{3}{2}$, $x \neq \dfrac{-2}{3}$.

EXAMPLE 12: If $y = 5x^2 - 3x + k$, for what values of k will the quadratic have two real solutions?

SOLUTION: $(-3)^2 - 4(5)k > 0 \Rightarrow 9 - 20k > 0 \Rightarrow k < \dfrac{9}{20}$.

9. Asymptotes

Rational functions in the form of $y = \dfrac{p(x)}{q(x)}$ may have vertical asymptotes, lines that the graph of the curve approach, but never cross. To find the vertical asymptotes, factor out any common factors of the numerator and denominator, reduce if possible, and then set the denominator equal to zero and solve.

Horizontal asymptotes are lines that the graph of the function approaches when x gets very large or very small. While you learn how to find these in calculus, a rule of thumb is that if the highest power of x is in the denominator, the horizontal asymptote is the line $y = 0$. If the highest power of x is both in the numerator and the denominator, the horizontal asymptote will be the line $y = \dfrac{\text{highest degree coefficient in the numerator}}{\text{highest degree coefficient in the denominator}}$.

If the highest power of x is in the numerator, there is no horizontal asymptote.

EXAMPLE 13: Find any vertical and horizontal asymptotes for the graph of $y = \dfrac{-x^2}{x^2 - x - 6}$.

SOLUTION: Since the denominator factors into $(x - 3)(x + 2)$, there are vertical asymptotes at $x = 3$ and $x = -2$.

Since there the highest power of x is 2 in both the numerator and the denominator, there is a horizontal asymptote at $y = \dfrac{-1}{1} = -1$.

This is confirmed by the graph on the next page. Note that the curve actually crosses its horizontal asymptote on the left side of the graph.

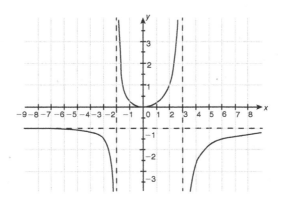

10. Inverses

No topic in math confuses students more than inverses. If a function is a rule that maps x to y, an inverse is a rule that brings y back to the original x. If a point (x, y) is a point on a function f, then the point (y, x) is on the inverse function f^{-1}. If a function is given in equation form, to find the inverse, replace all occurrences of x with y and all occurrences of y with x. If possible, solve for y. It is possible that the inverse of a function is not a function. Using the "horizontal-line test" on the original function f, will quickly determine whether or not f^{-1} is also a function.

EXAMPLE 14:   Find the inverse to $y = \dfrac{2x+5}{x-6}$.

SOLUTION: Inverse: $x = \dfrac{2y+5}{y-6}$

$$xy - 6x = 2y + 5 \Rightarrow xy - 2y = 6x + 5 \Rightarrow y = \frac{6x+5}{x-2}$$

EXAMPLE 15:   The graph of $y = \dfrac{-x^2}{x^2 - x - 6}$ is shown above. Is its inverse a function?

SOLUTION: No. At numerous locations, a horizontal line drawn will intersect the function in two locations.

11. Negative and Fractional Exponents

In calculus, you will be required to perform algebraic manipulations with negative exponents as well as fractional exponents. You should know the definition of a negative exponent: $x^{-n} = \dfrac{1}{x^n}, x \neq 0$. Note that negative exponents do not make expressions negative; they create fractions. Typically, expressions in multiple-choice answers will be written with positive exponents and students are required to eliminate the negative exponents. The definition of $x^{1/2} = \sqrt{x}$ and $x^{a/b} = \sqrt[b]{x^a} = \left(\sqrt[b]{x}\right)^a$.

EXAMPLE 16: Write with positive exponents

a) $-8x^{-2}$

b) $\left(27x^3\right)^{-2/3}$

c) $\left(x^{1/3}\right)\left(\dfrac{1}{2}x^{-1/2}\right) + \left(x^{1/2}+1\right)\left(\dfrac{1}{3}x^{-1/3}\right)$

SOLUTIONS:

a) $-8x^{-2} = \dfrac{-8}{x^2}$

b) $\left(27x^3\right)^{-2/3} = \dfrac{1}{\left(27x^3\right)^{2/3}} = \dfrac{1}{(27^{2/3})(x^2)} = \dfrac{1}{9x^2}$

c) $\dfrac{x^{1/3}}{2x^{1/2}} + \dfrac{x^{1/2}+1}{3x^{1/3}} = \dfrac{1}{2x^{1/6}} + \dfrac{x^{1/2}+1}{3x^{1/3}}$

12. Eliminating Complex Fractions

Calculus frequently uses complex fractions, which are fractions within fractions. Answers are never left with complex fractions and they must be eliminated. While there are other methods, the best way to accomplish this is to find the LCD (lowest common denominator) of all the fractions in the complex fraction. Multiply all terms by this LCD and you are left with a fraction that is magically no longer complex.

EXAMPLE 17:   Eliminate the complex fractions.

a) $\dfrac{1+\dfrac{1}{2x}}{1+\dfrac{1}{3x}}$

b) $\dfrac{\dfrac{1}{2}(2x+5)^{-2/3}}{\dfrac{-2}{3}}$

c) $\dfrac{(x-1)^{1/2}-\dfrac{x}{2(x-1)^{1/2}}}{x-1}$

SOLUTIONS: a) $\left(\dfrac{1+\dfrac{1}{2x}}{1+\dfrac{1}{3x}}\right)\left(\dfrac{6x}{6x}\right)=\dfrac{6x+3}{6x+2}$

b) $\left(\dfrac{\dfrac{1}{2}}{\dfrac{-2}{3}(2x+5)^{2/3}}\right)\dfrac{6}{6}=\dfrac{-3}{4(2x+5)^{2/3}}$

c) $\left(\dfrac{(x-1)^{1/2}-\dfrac{x}{2(x-1)^{1/2}}}{x-1}\right)\left(\dfrac{2(x-1)^{1/2}}{2(x-1)^{1/2}}\right)$

$\dfrac{2(x-1)-x}{2(x-1)^{3/2}}=\dfrac{x-2}{2(x-1)^{3/2}}$

13. Solving Fractional Equations

Algebra has taught you that to simplify an expression such as $\dfrac{12}{x+2}-\dfrac{4}{x}$, you need to find the LCD and multiply each fraction by this LCD in such a way that you have a common denominator. The LCD is $x(x+2)$, so you have

$$\left(\dfrac{12}{x+2}\right)\left(\dfrac{x}{x}\right)-\dfrac{4}{x}\left(\dfrac{x+2}{x+2}\right)=\dfrac{12x-4x-8}{x(x+2)}=\dfrac{8x-8}{x(x+2)}.$$

However, when you solve fractional equations (equations that involve fractions), you still find the LCD, but you multiply every term by the LCD. When you do that, all the fractions disappear, leaving you with an equation that is (hopefully) solvable.

EXAMPLE 18: Solve $\dfrac{12}{x+2} - \dfrac{4}{x} = 1$.

SOLUTION:
$$\dfrac{12}{x+2}(x)(x+2) - \dfrac{4}{x}(x)(x+2) = x(x+2)$$
$$12x - 4x - 8 = x^2 + 2x$$
$$x^2 - 6x + 8 = 0$$
$$(x - 2)(x - 4) = 0 \Rightarrow x = 2, 4$$

Note: Answers should be checked in the original equation because of the possibility of the denominator being zero.

$$x = 2 : \dfrac{12}{4} - \dfrac{4}{2} = 1 \Rightarrow 3 - 2 = 1 \Rightarrow 1 = 1$$

$$x = 4 : \dfrac{12}{6} - \dfrac{4}{4} = 1 \Rightarrow 2 - 1 = 1 \Rightarrow 1 = 1$$

14. Solving Absolute Value Equations

Absolute value equations crop up in calculus, especially in BC calculus. The definition of the absolute value function is a piecewise function: $f(x) = |x| = \begin{cases} x & \text{if } x \geq 0 \\ -x & \text{if } x < 0 \end{cases}$. To solve an absolute value equation, split the absolute value equation into two equations, one with a positive sign in front of the parentheses and the other with a negative sign in front of the parentheses and solve each equation.

EXAMPLE 19: Solve $|2x - 1| - x = 5$.

SOLUTION:

$(2x - 1) - x = 5$ $\qquad$ $-(2x - 1) - x = 5$
$2x - 1 - x = 5$ $\qquad$ $-2x + 1 - x = 5$

$x = 6$ $\qquad\qquad$ $x = \dfrac{-4}{3}$

EXAMPLE 20: Solve $\left| x^2 - x \right| = 2$.

SOLUTION:

$(x^2 - x) = 2$	$-(x^2 - x) = 2$
$x^2 - x - 2 = 0$	$-x^2 + x = 2$
$(x - 2)(x + 1) = 0$	$0 = x^2 - x + 2$
$x = 2, x = -1$	No real solution

15. Solving Inequalities

You may think that solving inequalities is just a matter of replacing the equal sign with an inequality sign. In reality, they can be more difficult and are fraught with dangers. In calculus, expect to solve a number of inequalities on a regular basis.

Solving inequalities is a simple matter if the inequalities are based on linear equations. They are solved exactly like linear equations, remembering that if you multiply or divide both sides by a negative number, the direction of the inequality sign must be reversed.

EXAMPLE 21: Solve $2x - 8 \le 6x + 2$.

SOLUTION:

$$-10 \le 4x \qquad\qquad\qquad -4x \le 10$$

$$\frac{-10}{4} \le x \text{ or } \frac{-5}{2} \le x \quad \text{OR} \quad x \ge \frac{-10}{4} \text{ or } x \ge \frac{-5}{2}$$

If the inequality is more complex than a linear function, it is advised to bring all terms to one side. Pretend for a minute it is an equation and solve. Then create a number line which determines whether the transformed inequality is positive or negative in the intervals created on the number line and choose the correct intervals according to the inequality, paying attention to whether the zeros are included or not.

EXAMPLE 22: Solve $x^2 - 3x > 18$.

SOLUTION:

$$x^2 - 3x - 18 > 0 \Rightarrow (x + 3)(x - 6) > 0$$

For $(x + 3)(x - 6) = 0$, $x = -3, x = 6$

$$+++++0-------0+++++++$$
$${-3}{6}$$

So $x < -3$ or $x > 6$ or $(-\infty, -3) \cup (6, \infty)$

EXAMPLE 23: Solve $|2x - 1| \le x + 4$

SOLUTION: $|2x - 1| \le x + 4 \Rightarrow |2x - 1| - x - 4 \le 0$

$2x - 1 - x - 4 = 0 \qquad -2x + 1 - x - 4 = 0$

$x = 5 \qquad\qquad x = -1$

$$+++++0-------0+++++++$$
$${-1}{5}$$

So $-1 \le x \le 5$ or $[-1, 5]$

If the inequality contains rational expressions, the method above for solving fractional equations by multiplying both sides of the equation by the LCD will not always work. It can lead to an incorrect solution because of the possibility of multiplying by zero. It is best, then, to bring all terms to one side, and set the numerator and the denominator equal to zero, solve each equation, and then create a sign chart showing intervals where the expression is positive or negative.

EXAMPLE 24: Solve $\dfrac{2x - 4}{x + 5} \le 0$.

SOLUTION: $2x - 4 = 0 \Rightarrow x = 2$

$x + 5 = 0 \Rightarrow x = -5$

$$+++++\infty-------0++++++$$
$${-5}{2}$$

So $-5 < x \le 2$ or $(-5, 2]$

Note that multiplying the original equation by $(x + 5)$ would lead to an incorrect solution.

16. Exponential Functions and Logs

Students must know that the definition of a logarithm is based on exponential equations. If $y = b^x$ then $x = \log_b y$. So when you are trying to find the value of $\log_2 32$, state that $\log_2 32 = x$ and $2^x = 32$. Therefore, $x = 5$.

If the base of a log statement is not specified, it is defined to be 10. The function $y = \log x$ has domain $(0, \infty)$ and range $(-\infty, \infty)$.

In calculus, we primarily use logs with base e, which are called natural logs (ln). So finding ln 5 is the same as solving the equation $e^x = 5$. Students should know that the value of $e = 2.71828\ldots$

There are three rules that students must keep in mind that will simplify problems involving logs and natural logs. These rules work with logs of any base including natural logs.

Rule 1. $\log (a \cdot b) \log a + \log b$

Rule 2. $\log\left(\dfrac{a}{b}\right) = \log a - \log b$

Rule 3. $\log a^b = b \log a$

EXAMPLE 25:   Find a) $\log_4 8$

b) $\ln \sqrt{e}$

c) $\log 250 - \log 2.5$

SOLUTIONS: a) $\log_4 8 = x$

$$4^x = 8 \Rightarrow 2^{2x} = 2^3$$

$$x = \frac{3}{2}$$

b) $\ln \sqrt{e} = x$

$$e^x = e^{1/2}$$

$$x = \frac{1}{2}$$

c) $\log 250 - \log 2.5 =$

$$\log\left(\frac{250}{2.5}\right)$$

$$\log 100 = 2$$

EXAMPLE 26: Solve for x: $\log_9\left(x^2 - x + 3\right) = \frac{1}{2}$.

SOLUTION: $x^2 - x + 3 = 9^{1/2} = 3$

$x(x-1) = 0$

$x = 0,\ x = 1$

EXAMPLE 27: Solve for x: $5^x = 20$.

SOLUTION:

$\log(5^x) = \log 20$ $\ln(5^x) = \ln 20$

$x \log 5 = \log 20$ or $x \ln 5 = \ln 20$

$x = \dfrac{\log 20}{\log 5} \approx 1.86$ $x = \dfrac{\ln 20}{\ln 5} \approx 1.86$

DIDYOUKNOW?

When numbers are very large or very small, they are expressed using a logarithmic scale. An example of this is the Richter scale, which measures the strength of an earthquake. An earthquake that measures 6 on the Richter scale is 10 times more destructive than one that measures 5 on the Richter scale. In 1994, an earthquake in Northridge California measured 6.7 on the Richter scale and caused about \$20 billion in damage. By contrast, the earthquake that struck Japan in 2011 measured 9.0 on the Richter scale. $\dfrac{10^9}{10^{6.7}} = 10^{2.3} \approx 200$. The Japanese earthquake was about 200 times more destructive than the California earthquake.

17. Right Angle Trigonometry

Trigonometry is an integral part of AP calculus. Students must know the basic trigonometric function definitions in terms of opposite, adjacent, and hypotenuse, as well as the definitions if the angle is in standard position.

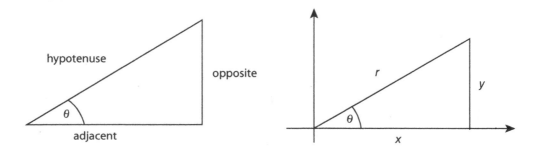

Given a right triangle with one of the angles named θ, and the sides of the triangle relative to θ named opposite, adjacent, and hypotenuse (graph on the left), we define the six trigonometric functions to be:

Basic Trigonometry Definitions

the sine function: $\sin\theta = \dfrac{\text{opposite}}{\text{hypotenuse}}$ 　　 the cosecant function: $\csc\theta = \dfrac{\text{hypotenuse}}{\text{opposite}}$

the cosine function: $\cos\theta = \dfrac{\text{adjacent}}{\text{hypotenuse}}$ 　　 the secant function: $\sec\theta = \dfrac{\text{hypotenuse}}{\text{adjacent}}$

the tangent function: $\tan\theta = \dfrac{\text{opposite}}{\text{adjacent}}$ 　　 the cotangent function: $\cot\theta = \dfrac{\text{adjacent}}{\text{opposite}}$

Given a right triangle with one of the angles named θ with θ in standard position, and the sides of the triangle relative to θ named x, y, and r. (see the graph on the right above), we define the six trigonometric functions to be:

the sine function: $\sin\theta = \dfrac{y}{r}$ 　　 the cosecant function: $\csc\theta = \dfrac{r}{y}$

the cosine function: $\cos\theta = \dfrac{x}{r}$ 　　 the secant function: $\sec\theta = \dfrac{r}{x}$

the tangent function: $\tan\theta = \dfrac{y}{x}$ 　　 the cotangent function: $\cot\theta = \dfrac{x}{y}$

The Pythagorean theorem ties these variables together: $x^2 + y^2 = r^2$. Since the Pythagorean theorem is important in right triangles, students should recognize right triangles with integer sides: 3-4-5, 5-12-13, 8-15-17, 7-24-25. Remember that any multiples of these sides are also sides of a right triangle.

Also vital to master are the signs of the trigonometric functions in the four quadrants. A good way to remember this is

$$A - S - T - C$$

<u>A</u>ll trig functions are positive in the 1st quadrant,

<u>S</u>in is positive in the 2nd quadrant,

<u>T</u>an is positive in the 3rd quadrant, and

<u>C</u>os is positive in the 4th quadrant.

EXAMPLE 28: Let $P(-8, 6)$ be a point on the terminal side of θ. Find the 6 trigonometric functions of θ.

SOLUTION: $x = -8, y = 6, r = 10$

$$\sin\theta = \frac{3}{5} \qquad\qquad \csc\theta = \frac{5}{3}$$

$$\cos\theta = -\frac{4}{5} \qquad\qquad \sec\theta = -\frac{5}{4}$$

$$\tan\theta = -\frac{3}{4} \qquad\qquad \cot\theta = -\frac{4}{3}$$

EXAMPLE 29: If $\cos\theta = \frac{2}{3}$, θ in quadrant IV, find $\sin\theta$ and $\tan\theta$

SOLUTION: $x = 2, r = 3, y = -\sqrt{5}$

$$\sin\theta = -\frac{\sqrt{5}}{3}, \tan\theta = -\frac{\sqrt{5}}{2}$$

18. Special Angles

Students must be able to find trigonometric functions of quadrant angles (0, 90°, 180°, 270°) and special angles, those based on the 30°-60°-90° and 45°-45°-90° triangles.

First, for most calculus problems, angles are given and found in radians. Students must know how to convert degrees to radians and vice-versa. The relationship is 2π radians = 360° or π radians = 180°. Angles are assumed to be in radians so when an angle of $\dfrac{\pi}{3}$ is given, it is in radians. However, a student should be able to picture this angle as $\dfrac{180°}{3} = 60°$. It may be easier to think of angles in degrees than radians, but realize that unless specified, angle measurement is written in radians. For instance, $\sin^{-1}\left(\dfrac{1}{2}\right) = \dfrac{\pi}{6}$.

The trig functions of quadrant angles $\left(0,\ 90°, 180°, 270°\ \text{or}\ 0,\ \dfrac{\pi}{2}, \pi, \dfrac{3\pi}{2}\right)$ can quickly be found. Choose a point along the angle and realize that r is the distance from the origin to that point and always positive. Then use the definitions of the trigonometric functions.

θ	point	x	y	r	$\sin \theta$	$\cos \theta$	$\tan \theta$
0	(1,0)	1	0	1	0	1	0
$\dfrac{\pi}{2}$ or 90°	(0,1)	0	1	1	1	0	Does not exist
π or 180°	(−1,0)	−1	0	1	0	−1	0
$\dfrac{3\pi}{2}$ or 270°	(0,−1)	0	−1	1	−1	0	Does not exist

The table above goes hand-in-hand by the shape of the two toolbox functions: $y = \sin x$ and $y = \cos x$.

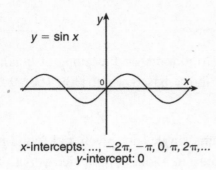

$y = \sin x$

x-intercepts: ..., -2π, $-\pi$, 0, π, 2π,...
y-intercept: 0

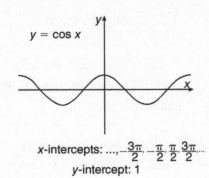

$y = \cos x$

x-intercepts: ..., $-\dfrac{3\pi}{2}$, $-\dfrac{\pi}{2}$, $\dfrac{\pi}{2}$, $\dfrac{3\pi}{2}$,...
y-intercept: 1

You must know the relationship of sides in both

$$30° - 60° - 90° \left(\frac{\pi}{6}, \frac{\pi}{3}, \frac{\pi}{2}\right) \text{ and } 45° - 45° - 90° \left(\frac{\pi}{4}, \frac{\pi}{4}, \frac{\pi}{2}\right) \text{ triangles.}$$

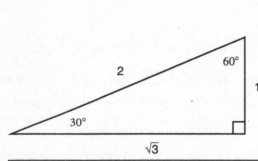

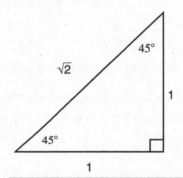

In a $30° - 60° - 90° \left(\dfrac{\pi}{6}, \dfrac{\pi}{3}, \dfrac{\pi}{2}\right)$ triangle, the ratio of sides is $1 : \sqrt{3} : 2$.

In a $45° - 45° - 90° \left(\dfrac{\pi}{4}, \dfrac{\pi}{4}, \dfrac{\pi}{2}\right)$ triangle, the ratio of sides is $1 : 1 : \sqrt{2}$.

θ	$\sin \theta$	$\cos \theta$	$\tan \theta$
$30° \left(\text{or } \dfrac{\pi}{6}\right)$	$\dfrac{1}{2}$	$\dfrac{\sqrt{3}}{2}$	$\dfrac{\sqrt{3}}{3}$
$45° \left(\text{or } \dfrac{\pi}{4}\right)$	$\dfrac{\sqrt{2}}{2}$	$\dfrac{\sqrt{2}}{2}$	1
$60° \left(\text{or } \dfrac{\pi}{3}\right)$	$\dfrac{\sqrt{3}}{2}$	$\dfrac{1}{2}$	$\sqrt{3}$

Special angles are any multiple of 30° $\left(\dfrac{\pi}{6}\right)$ or 45° $\left(\dfrac{\pi}{2}\right)$. To find trigonometric functions of any of these angles, draw them and find the reference angle (the angle created with the x-axis). This will create one of the triangles above and trigonometric functions can be found. Remember to include the sign based on the quadrant of the angle.

EXAMPLE 30: Find the exact value of the following:

a) $4\sin\dfrac{2\pi}{3} - 8\cos\dfrac{7\pi}{6}$

b) $\left(2\cos\pi - 5\tan\dfrac{7\pi}{4}\right)^2$

SOLUTIONS: a) $4\sin 120° - 8\cos 210°$

120° is in quadrant II with reference angle 60°.

210° is in quadrant III with reference angle 30°.

$4\left(\dfrac{\sqrt{3}}{2}\right) - 8\left(\dfrac{-\sqrt{3}}{2}\right) = 6\sqrt{3}$

b) $(2\cos180° - 5\tan315°)^2$

180° is a quadrant angle

315° is in quadrant IV with reference angle 45°

$[2(-1) - 5(-1)]^2 = 9$

19. Trigonometric Identities

Trigonometric identities are equalities involving trigonometric functions that are true for all values of the given angles. While you are not asked these identities specifically in calculus, knowing them can make some problems easier. The following chart gives the major trigonometric identities; those with an asterisk indicate identities that are more likely to appear.

Fundamental Trig Identities
* $\csc x = \dfrac{1}{\sin x}$
* $\sec x = \dfrac{1}{\cos x}$
* $\cot x = \dfrac{1}{\tan x}$
* $\tan x = \dfrac{\sin x}{\cos x}$
$\cot x = \dfrac{\cos x}{\sin x}$
* $\sin^2 x + \cos^2 x = 1$
$1 + \tan^2 x = \sec^2 x$
$1 + \cot^2 x = \sec^2 x$

Sum Identities
$\sin(A + B) = \sin A \cos B + \cos A \sin B$
$\cos(A + B) = \cos A \cos B - \sin A \sin B$

Double Angle Identities
$\sin(2x) = 2\sin x \cos x$
$\cos(2x) = \cos^2 x - \sin^2 x = 1 - 2\sin^2 x = 2\cos^2 x - 1$

20. Solving Trigonometric Equations or Inequalities

Trigonometric equations are equations using trigonometric functions. Typically, they have many (or an infinite number of) solutions, so usually they are solved within a specific domain. Without calculators, answers are either quadrant angles or special angles, and again, they must be expressed in radians.

For trigonometric inequalities, set both the numerator and the denominator equal to zero and solve. Make a sign chart with all these values included and examine the sign of the expression in the intervals. Basic knowledge of the sine and cosine curve is invaluable.

EXAMPLE 31: Solve for x on $[0, 2\pi)$: $x \cos x = 2 \cos x$.

SOLUTION: Do not divide by $\cos x$ as you will lose solutions.

$$x \cos x - 2 \cos x = 0$$

$$\cos x \, (x - 2) = 0$$

$$\cos x = 0 \Rightarrow x = \frac{\pi}{2}, \frac{3\pi}{2}$$

$$x - 2 = 0 \Rightarrow x = 2$$

The solutions are: $x = 2, \dfrac{\pi}{2}, \dfrac{3\pi}{2}$

This problem shows you the need to work in radians. Saying $x = 90°$ would make no sense.

EXAMPLE 32: Solve for x on $[0, 2\pi)$: $\tan x + \sin^2 x = 2 - \cos^2 x$.

SOLUTION: $\tan x + \sin^2 x + \cos^2 x = 2$

Using identity : $\tan x + 1 = 2$

$\tan x = 1$

$$x = \frac{\pi}{4}, \frac{5\pi}{4}$$

The tangent function is positive in quadrants I and III.

EXAMPLE 33: Solve for x on $[0, 2\pi)$ $\dfrac{2\cos x - 1}{\sin^2 x} > 0$.

SOLUTION:

$$2\cos x = 1 \Rightarrow \cos x = \frac{1}{2} \Rightarrow x = \frac{\pi}{3}, \frac{5\pi}{3}$$

$$\sin^2 x = 0 \Rightarrow \sin x = 0 \Rightarrow x = 0, \pi$$

$$\begin{array}{ccccc} +\!+\!+\!+\!+\!+\!+\!+\!+0 & -\!-\!-\!-\!-\!-\!-\infty & -\!-\!-\!-\!-\!-\!-0 & +\!+\!+\!+\!+\!+\!+\!+\!0 \\ \hline 0 & \frac{\pi}{3} & \pi & \frac{5\pi}{3} & 2\pi \end{array}$$

Answer: $\left(0, \dfrac{\pi}{3}\right) \cup \left(\dfrac{5\pi}{3}, 2\pi\right)$

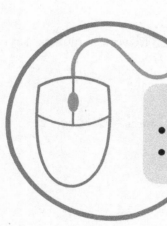

Time for a quiz
- Review strategies in Chapter 2
- Take Quiz 1 at the REA Study Center

(www.rea.com/studycenter)

AP Calculus AB

Basic Calculus Concepts and Limits

Introduction

On the AP exam, there is no requirement to know any of the fascinating evolution of calculus. However, many students do not even know what calculus is; they just know it is a math course.

I think of calculus as the study of change. In courses before calculus, the value of x is fixed. You were told that $x = 5$, the hypotenuse of the triangle is 10, or the height of the water in the tank is 4 feet. You knew that, at least for that problem, those values would never change.

But in calculus, you might be told the value of x, but you are also told that x is changing. That opens up a brand new world. Think of all the aspects of yourself that are changing. You are getting taller, your hair and fingernails are growing, your weight might be increasing or decreasing, and blood is not stationary but moving throughout your body. And because so many things in the world are changing, calculus is a course for which many real-life problems are pertinent, especially in science, economics, and engineering.

There are two basic branches of calculus. The first is called **differential calculus** and the problem that it studies is finding the tangent line to a graph at a point. In the figure below, line L is tangent to the graph of $f(x)$ at point P. Chapters 5 and 6 of this book are concerned with the issue of the tangent line problem and the doors this problem opens.

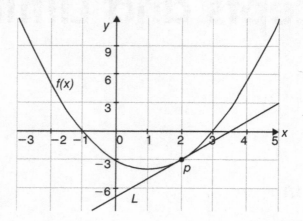

The second branch of calculus is called **integral calculus** and the problem that it studies is finding the area under a curve between two vertical lines. In the figure below, the shaded area between the graph of $f(x)$, the lines $x = -2$, $x = 1$, and the x-axis is 26.25. Chapters 7 and 8 are concerned with the issue of the area problem and the doors that this problem opens.

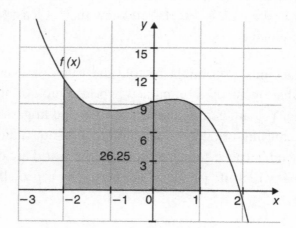

DIDYOUKNOW?

Sir Isaac Newton (1642–1727) actually discovered calculus between 1665 and 1667 after his university closed due to an outbreak of the plague. Newton was only 22 at the time, and he preferred not to publish his discoveries. Meanwhile, in Germany, Gottfried Wilhelm Leibniz discovered calculus independently and was very open with his findings. This led to a bitter dispute between the two mathematicians later known as the "Great Sulk." Today it is well known that both men discovered calculus independently of the other, Leibniz about 8 years after Newton.

Graphical Approach to Limits

Overview: The backbone of calculus is limits. While only about 5% of the AP exam actually tests limits directly, all of the concepts of differential and integral concept are based on limits. In this section, we will find limits when we are given graphs of functions.

A limit is a boundary. A speed limit is the maximum speed you can travel without being in danger of receiving a ticket. A student who knows his limits with his parents knows exactly how far he can push his parents without them coming down on him. In real life, sometimes a limit can be somewhat hazy. For instance, I might not be sure of my limit of tolerance of heat before I turn the air conditioner on. But I do know that 70 degrees is on one side of that limit and 90 degrees is on the other side. The study of limits is crucial to calculus, dictating that most calculus courses begin with a study of them.

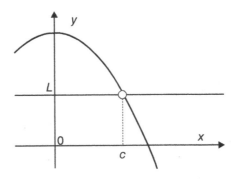

The notation for limits is typically: $\lim_{x \to c} f(x) = L$. What this says is: as the value of x gets close to the value of c, the value of y gets close to some constant L. The figure above demonstrates this: The closer the x-value is to c, the closer the y-value is to L.

Whether the y-value ever reaches L is irrelevant. We are only interested in the value that the y-value is approaching. In fact, in the figure, the y-value does not ever reach L at $x = c$ as there is a hole in the graph. It doesn't matter. The limit still exists.

In order for a limit to exist, we say that $\lim_{x \to c^-} f(x)$ must equal $\lim_{x \to c^+} f(x)$. What this says in words is: as x approaches c on the left side of c and as x approaches c on the right side of c, the y-values must be approaching the same number. If they do not, we say that $\lim_{x \to c} f(x)$ does not exist.

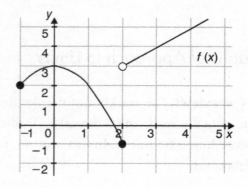

EXAMPLE 1:   Given the graph of $f(x)$, find $\lim_{x \to 2} f(x)$.

SOLUTION: $\lim_{x \to 2^-} f(x) = -1$ and $\lim_{x \to 2^+} f(x) = 3$

Since $-1 \neq 3$, $\lim_{x \to 2} f(x)$ does not exist.

Note that from the given graph, it is clear that $f(2) = -1$.
That does not affect the calculation of the limit.

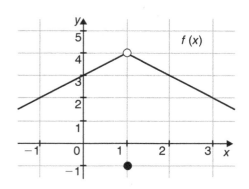

EXAMPLE 2: Given the graph of $f(x)$, find

a. $\lim_{x \to 1} f(x)$

b. $f(1)$

c. $\lim_{x \to -1} f(x)$.

SOLUTIONS: a. $\lim_{x \to 1^-} f(x) = 4$ and $\lim_{x \to 1^+} f(x) = 4$.

So $\lim_{x \to 1} f(x) = 4$. The open dot at $(1, 4)$ does not matter.

b. $f(1) = -1$ as there is a closed dot at $(1, -1)$. However, this has absolutely nothing to do with $\lim_{x \to 1} f(x)$.

c. $\lim_{x \to -1^-} f(x) = 2$ and $\lim_{x \to -1^+} f(x) = 2$. So $\lim_{x \to -1} f(x) = 2$. Also, $f(-1) = 2$. This is an example of the graph being continuous (covered later in this chapter) at $x = -1$ and whenever graphs are continuous at $x = c$, $\lim_{x \to c^-} f(x) = \lim_{x \to c^+} f(x) = \lim_{x \to c} f(x) = f(c)$.

Graphs that have vertical asymptotes have non-existent limits as the value of x approaches the vertical asymptote. If the graphs on both sides of the vertical asymptote are going up, it is convenient to say that $\lim_{x \to c} f(x) = \infty$. If the graphs on both sides of the vertical asymptote are going down, it is convenient to say that $\lim_{x \to c} f(x) = -\infty$. But, if on one side of the vertical asymptote the graph is going up and on the other side the graph is going down, we cannot make any claim other than the limit doesn't exist.

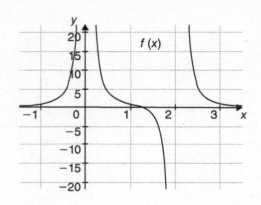

EXAMPLE 3: Given the graph of $f(x)$, above, find

a. $\lim\limits_{x \to 0} f(x)$

b. $\lim\limits_{x \to 2} f(x)$.

SOLUTIONS: a. $\lim\limits_{x \to 0^-} f(x) = \infty$ and $\lim\limits_{x \to 0^+} f(x) = \infty$

So $\lim\limits_{x \to 0} f(x) = \infty$.

b. $\lim\limits_{x \to 2^-} f(x) = -\infty$ and $\lim\limits_{x \to 2^+} f(x) = \infty$

So $\lim\limits_{x \to 2} f(x)$ does not exist.

TEST TIP

It is important to realize that saying $\lim\limits_{x \to c} f(x) = \infty$ is saying that the limit does not exist. A limit being infinite is saying that there is no limit. We write $\lim\limits_{x \to c} f(x) = \infty$ as a convenience to indicate that the graph of $f(x)$ goes up forever on both sides of c. Using the graph in Example 3, if you were asked on the AP exam for what values of c, the $\lim\limits_{x \to c} f(x)$ does not exist, your answers would be both $c = 0$ and $c = 2$, albeit for different reasons.

We are also interested in the limits: $\lim_{x \to \infty} f(x)$ and $\lim_{x \to -\infty} f(x)$. The meaning of $\lim_{x \to \infty} f(x)$ is what happens to y as x gets infinitely further to the right and the meaning of $\lim_{x \to -\infty} f(x)$ is what happens to y as x gets infinitely further to the left. There are only 4 possible situations that can occur graphically to illustrate these limits:

1. $\lim_{x \to \infty} f(x) = \infty$. This means that the further to the right, the higher the graph gets. Since infinity does not exist as a number, saying that $\lim_{x \to \infty} f(x) = \infty$ means that the limit does not exist.

2. $\lim_{x \to \infty} f(x) = -\infty$. This means that the further to the right, the lower the graph gets. Since infinity does not exist as a number, saying that $\lim_{x \to \infty} f(x) = -\infty$ means that the limit does not exist.

3. $\lim_{x \to \infty} f(x) = L$. This means that the further to the right, the closer a curve gets to some constant L. There are two ways this can happen. One way is becoming asymptotic to the line $y = L$ and the other way is to oscillate about L but constantly getting closer to it. Saying that a function has a horizontal asymptote at $y = L$ is saying that either $\lim_{x \to \infty} f(x) = L$ or $\lim_{x \to -\infty} f(x) = L$ or both.

4. $\lim_{x \to \infty} f(x)$ does not exist. While situations 1 and 2 have limits that do not exist, these graphs are either going up or down as x gets smaller or larger. In this situation, the graphs usually oscillate between two values or oscillate between values that are getting further away from each other.

DIDYOU**KNOW?**

"The infinite! No other question has ever moved so profoundly the spirit of man."—Mathematician David Hilbert

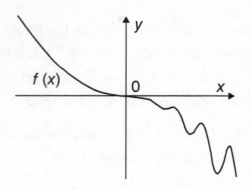

EXAMPLE 4:   Using the figure above, find

a. $\lim\limits_{x \to -\infty} f(x)$

b. $\lim\limits_{x \to \infty} f(x)$

SOLUTIONS: This example illustrates the first two situations as described above. Based on the information in this graph, we can make the following statements:

a. $\lim\limits_{x \to -\infty} f(x) = \infty$ ... the further to the left, the higher the curve gets.

b. $\lim\limits_{x \to \infty} f(x) = -\infty$ ... the further to the right, the lower the curve gets. It oscillates and comes back up, but based on the diagram, the curve gets lower.

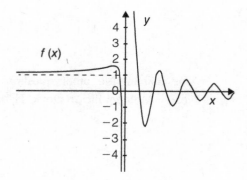

EXAMPLE 5:   Using the figure above, find

a. $\lim\limits_{x \to -\infty} f(x)$

b. $\lim\limits_{x \to \infty} f(x)$.

SOLUTIONS: This example illustrates the 3rd situation as described above. Based on the information in this graph, we can make the following statements:

a. $\lim_{x \to -\infty} f(x) = 1$. There is a horizontal asymptote along the line $y = 1$ and the further to the left, the closer we get to that line. Note that the graph of f passes through the line $y = 1$. That does not change the fact that $\lim_{x \to -\infty} f(x) = 1$. While functions can never cross a vertical asymptote, they can pass through horizontal asymptotes.

b. $\lim_{x \to \infty} f(x) = 0$. The further to the right, the closer the curve gets to the x-axis. It oscillates above and below this line but always getting closer. The fact that it crosses the x-axis an infinite number of times doesn't change the fact that $\lim_{x \to \infty} f(x) = 0$.

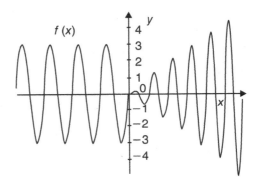

EXAMPLE 6: Using the figure above, find

a. $\lim_{x \to -\infty} f(x)$

b. $\lim_{x \to \infty} f(x)$.

SOLUTIONS: This example illustrates the 4th situation as described above. Based on the information in this graph, we can make the following statements:

a. $\lim_{x \to -\infty} f(x)$ does not exist. The graph oscillates between -3 and 3, not getting closer to any single y-value.

b. $\lim_{x \to \infty} f(x)$ does not exist. The graph oscillates between values that are getting further away from each other.

Algebraic Approach to Limits

Overview: While finding limits, as x approaches a value or infinity, looking at the graph makes the solution easy to find. However, you are required to be either given the graph or have a graphing utility and be able to view it in a window that shows its important behavior. Far more representative of AP exam problems are those that ask for limits when you are given the function in algebraic form.

For many limit problems, there is a 3-step process that, given a function $f(x)$, will determine if the limit of $f(x)$ as x approaches c exists and if so, what it is. That is finding: $\lim_{x \to c} f(x)$.

1. Evaluate f at c. That is, find $f(c)$. If this exists, that is the limit.

2. If step 1 gives an expression that does not exist (a zero in the denominator), attempt to factor both the numerator and the denominator of $f(x)$, cancel common terms to simplify the expression, and go back to step 1.

3. If after factoring, you still get an expression that does not exist (a zero in the denominator), $\lim_{x \to c} f(x)$ does not exist. To determine whether $\lim_{x \to c} f(x) = \infty$ or $\lim_{x \to c} f(x) = -\infty$, you need to determine the sign of $\lim_{x \to c^-} f(x)$ and $\lim_{x \to c^+} f(x)$. They will both be infinite – the only question is whether their signs are the same.

EXAMPLE 7: Find $\lim_{x \to 4} x^2 - 3x - 2$.

SOLUTION: $\lim_{x \to 4} x^2 - 3x - 2 = (4)^2 - 3(4) - 2 = 2$.

EXAMPLE 8: Find $\lim_{x \to r} \dfrac{(2x - r)^2}{x^2 + 4x + r}$.

SOLUTION: $\lim_{x \to r} \dfrac{(2x - r)^2}{x^2 + 4x + r} = \dfrac{(2r - r)^2}{r^2 + 4r + r} = \dfrac{r^2}{r^2 + 5r} = \dfrac{r}{r + 5}$.

EXAMPLE 9: Find $\lim\limits_{x \to 3} \dfrac{x^2 - x - 6}{x - 3}$.

SOLUTION: $\lim\limits_{x \to 3} \dfrac{x^2 - x - 6}{x - 3} = \dfrac{3^2 - 3 - 6}{3 - 3} = \dfrac{0}{0}$

$\lim\limits_{x \to 3} \dfrac{(x-3)(x+2)}{x-3} = \lim\limits_{x \to 3}(x+2) = 3 + 2 = 5$

EXAMPLE 10: Find $\lim\limits_{x \to -1} \dfrac{x^3 + 1}{x + 1}$.

SOLUTION: $\lim\limits_{x \to -1} \dfrac{x^3 + 1}{x + 1} = \dfrac{-1 + 1}{-1 + 1} = \dfrac{0}{0}$

$\lim\limits_{x \to -1} \dfrac{(x+1)(x^2 - x + 1)}{x + 1} = \lim\limits_{x \to -1}(x^2 - x + 1) = 1 + 1 + 1 = 3$

TEST TIP

The type of factoring in Example 10 involving the sum of cubes, while not common on the AP exam, is fair game. If your factoring skills are weak and if you can use the graphing calculator for this type of problem, you can ascertain the limit by plugging in a number very close to the value of c for which you want the limit. While not conclusive proof that the limit requested in Example 10 is 3, the calculator screens shown below are strong evidence of that fact.

```
Plot1  Plot2  Plot3
\Y₁=(X³+1)/(X+1)
```

```
Y₁(−.9999)
         2.99970001
Y₁(−1.0001)
         3.00030001
```

EXAMPLE 11: Find $\lim\limits_{x \to 2} \dfrac{x - 5}{(x - 2)^2}$.

SOLUTION: $\lim\limits_{x \to 2} \dfrac{x - 5}{(x - 2)^2} = \dfrac{2 - 5}{(2 - 2)^2} = \dfrac{-3}{0}$, so $\lim\limits_{x \to 2} \dfrac{x - 5}{(x - 2)^2}$ does not exist.

If more information is needed for this answer, we break up the problem into left-hand and right-hand limits.

$\lim\limits_{x \to 2^-} \dfrac{x-5}{(x-2)^2}$: Choose a number for x close to 2 on the left-hand side. Use 1.9. Plug 1.9 into the expression and calculate only its sign (because its value is infinite). $\lim\limits_{x \to 2^-} \dfrac{x-5}{(x-2)^2} : \dfrac{-}{+}\infty = -\infty$.

$\lim\limits_{x \to 2^+} \dfrac{x-5}{(x-2)^2}$: Choose a number for x close to 2 on the right-hand side. Use 2.1. Plug 2.1 into the expression and calculate only its sign (because its value is infinite). $\lim\limits_{x \to 2^+} \dfrac{x-5}{(x-2)^2} : \dfrac{-}{+}\infty = -\infty$.

Since the signs are the same, we can say $\lim\limits_{x \to 2} \dfrac{x-5}{(x-2)^2} = -\infty$. But remember, this limit doesn't exist because $-\infty$ is not a number.

This solution is confirmed graphically:

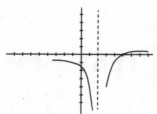

EXAMPLE 12:   Find $\lim\limits_{x \to -5} \dfrac{x^2+2x-15}{x^2+10x+25}$.

SOLUTION: $\lim\limits_{x \to -5} \dfrac{x^2+2x-15}{x^2+10x+25} = \dfrac{25-10-15}{25-50+25} = \dfrac{0}{0}$

$\lim\limits_{x \to -5} \dfrac{(x+5)(x-3)}{(x+5)(x+5)} = \lim\limits_{x \to -5} \dfrac{x-3}{x+5} = \dfrac{-8}{0}$

$\lim\limits_{x \to -5^-} \dfrac{x-3}{x+5} = \dfrac{-}{-}\infty = \infty$

$\lim\limits_{x \to -5^+} \dfrac{x-3}{x+5} = \dfrac{-}{+}\infty = -\infty$

So $\lim\limits_{x \to -5} \dfrac{x^2+2x-15}{x^2+10x+25}$ does not exist.

EXAMPLE 13: For the piecewise function $f(x) = \begin{cases} 2x^2 - 5x + 6, & x \geq 2 \\ 3 + \cos(x-2), & x < 2 \end{cases}$,

find $\lim\limits_{x \to 2} f(x)$.

SOLUTION: For piecewise functions, the same rules apply, but you must split the function into two pieces and determine if the left-hand and right-hand limits exist and whether they are the same.

$$\lim_{x \to 2^-} f(x) = 3 + \cos 0 = 3 + 1 = 4$$

$$\lim_{x \to 2^+} f(x) = 8 - 10 + 6 = 4$$

So $\lim\limits_{x \to 2} f(x) = 4$

This is confirmed graphically:

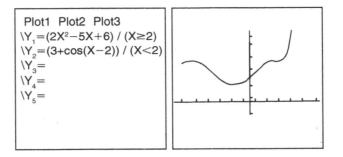

EXAMPLE 14: If $f(x) = \begin{cases} \dfrac{|x-4|}{x-4}, & x \neq 4 \\ 1, & x = 4 \end{cases}$, find $\lim\limits_{x \to 4} f(x)$.

SOLUTION: Rewrite $f(x)$ as a piecewise function:

$$f(x) = \begin{cases} \dfrac{(x-4)}{x-4}, & x > 4 \\ \dfrac{-(x-4)}{x-4}, & x < 4 \\ 1, & x = 4 \end{cases} = \begin{cases} 1, & x > 4 \\ -1, & x < 4 \\ 1, & x = 4 \end{cases}$$

$$\lim_{x \to 4^-} f(x) = -1 \text{ and } \lim_{x \to 4^+} f(x) = 1$$

So $\lim\limits_{x \to 4} f(x)$ does not exist. Note that $f(4)$ is irrelevant to the problem. The graph of the function is as follows:

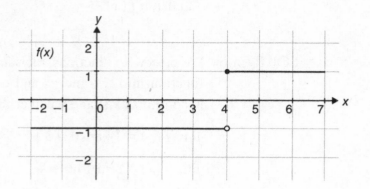

EXAMPLE 15: Find $\lim\limits_{x \to 0} \dfrac{\sqrt{x+6}-\sqrt{6}}{x}$.

SOLUTION: Evaluating at $x = 0$ gives $\dfrac{0}{0}$. The indeterminate $\dfrac{0}{0}$ is an indication that the limit actually exists. But there is no obvious factoring that can be done here. The solution is to multiply both the numerator and the denominator by the conjugate of $\sqrt{x+6}-\sqrt{6}$ which is $\sqrt{x+6}+\sqrt{6}$. Good things happen!

$$\lim_{x \to 0}\left(\frac{\sqrt{x+6}-\sqrt{6}}{x}\right)\left(\frac{\sqrt{x+6}+\sqrt{6}}{\sqrt{x+6}+\sqrt{6}}\right) = \lim_{x \to 0}\frac{x+6-6}{x\left(\sqrt{x+6}+\sqrt{6}\right)} = \frac{1}{2\sqrt{6}}$$

This is confirmed using the calculator. Choosing a number close to zero yields a value close to $\dfrac{1}{2\sqrt{6}}$.

```
Plot1  Plot2  Plot3
 Y₁=(√(X+6)−√(6))/ X
\Y₂=
\Y₃=
\Y₄=
\Y₅=
\Y₆=
```

```
Y₁(.0001)
        .204123295
Y₁(−.0001)
        .204124996
1/(2√(6))
        .2041241452
```

TEST TIP

It is not necessary to rationalize fractions in the AP exam. While your teacher might require you to change $\dfrac{1}{2\sqrt{6}}$ to $\dfrac{\sqrt{6}}{12}$, it is not required (nor recommended) on the AP exam. If you take the time and do so, it can cost you points if you make an error. Of course, on multiple-choice problems, either answer could appear.

EXAMPLE 16: (Calculator Active) Find $\displaystyle\lim_{x\to 0}\frac{\sin x}{x}$.

SOLUTION: Evaluating at $x = 0$ gives $\dfrac{0}{0}$ and there is no factoring/canceling that you can do. One would think that means the limit does not exist, but not necessarily. When you plug in and get $\dfrac{0}{0}$, that is an indication that the limit may very well exist. To find out, you need to learn the BC topic L'Hospital's Rule which is covered in Chapter 9. For AB students, the only way you can solve this is to look at the problem graphically. This type of problem would never appear on the non-calculator section of the AB exam, but it is fair game for the calculator-active section.

It appears that the limit is 1 even though the graph is undefined at $x = 0$. Evaluating expressions close to 0 seems to confirm that fact.

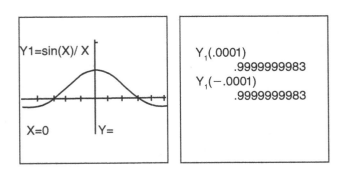

Finding limits of functions as x approaches infinity or negative infinity $\left(\displaystyle\lim_{x\to\infty} f(x) \text{ or } \lim_{x\to-\infty} f(x)\right)$ can be broken down into several simple rules that will work most of the time:

- First, write the function as a fraction.

- If the highest power of x in the function appears in the denominator (bottom-heavy), $\lim_{x \to \infty} f(x) = 0$ and $\lim_{x \to -\infty} f(x) = 0$.

- If the highest power of x in the function appears in the numerator (top-heavy), $\lim_{x \to \infty} f(x)$ and $\lim_{x \to -\infty} f(x)$ do not exist. To find out whether the limit is either positive or negative infinity, examine the sign of the expression when very large numbers are plugged in.

- If the highest power of x appears both in the numerator and the denominator (powers equal), $\lim_{x \to \infty} f(x)$ and $\lim_{x \to -\infty} f(x)$ will be the ratio of the coefficients of the highest power terms.

EXAMPLE 17: Find $\lim_{x \to \infty} \dfrac{20x^2 + 10x + 5}{1 - x^3}$.

SOLUTION: This is a "bottom-heavy" expression and $\lim_{x \to \infty} \dfrac{20x^2 + 10x + 5}{1 - x^3} = 0$. The coefficients do not matter. Ultimately, the denominator is more "powerful" than the numerator.

EXAMPLE 18: Find $\lim_{x \to -\infty} \dfrac{x^2 - 4x - 3}{1 - 5x}$.

SOLUTION: This is a "top-heavy expression" so $\lim_{x \to -\infty} \dfrac{x^2 - 4x - 3}{1 - 5x}$ does not exist. More detail can be given as if x were a number like -100, the numerator would be positive and the denominator would be positive as well. We can also say that $\lim_{x \to -\infty} \dfrac{x^2 - 4x - 3}{1 - 5x} = \infty$.

EXAMPLE 19: Find $\lim\limits_{x \to -\infty} \dfrac{4x^2 - 7x + 2}{1 - x - 3x^2}$.

SOLUTION: This is a "powers equal" expression so $\lim\limits_{x \to \infty} \dfrac{4x^2 - 7x + 2}{1 - x - 3x^2} = \dfrac{4}{-3} = \dfrac{-4}{3}$.

There are some problems that don't quite fit the rules shown above.

EXAMPLE 20: Find a) $\lim\limits_{x \to \infty} \dfrac{\sqrt{4x^2 + 3x + 5}}{2x + 7}$

b) $\lim\limits_{x \to -\infty} \dfrac{\sqrt{4x^2 + 3x + 5}}{2x + 7}$.

SOLUTIONS: a) We can ignore everything but the highest-power term under the square root function so the expression becomes $\lim\limits_{x \to \infty} \dfrac{\sqrt{4x^2}}{2x} = \lim\limits_{x \to \infty} \dfrac{2x}{2x} = 1$.

b) You might think that $\lim\limits_{x \to -\infty} \dfrac{\sqrt{4x^2 + 3x + 5}}{2x + 7}$ would be handled exactly as answer a), but realize that as x approaches negative infinity, the numerator will always be positive while the denominator would be negative. So $\lim\limits_{x \to -\infty} \dfrac{\sqrt{4x^2 + 3x + 5}}{2x + 7} = -1$. This is confirmed graphically.

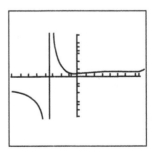

The rules above for finding limits at infinity don't apply for all functions, but a little common sense will help you find these limits for such functions.

TEST TIP

Avoid using "jargon" on the AP exam. While terms like "bottom heavy" can help you to learn concepts, to a test reader, they may have little meaning. If you are asked for a justification, explain yourself. Also never use any calculator jargon in the AP exam. A statement like "I put the function in YI and used the NDERIV function" has no meaning to an AP reader.

EXAMPLE 21: Find a) $\lim\limits_{x \to \infty} \dfrac{x^{10}}{2^x}$

b) $\lim\limits_{x \to -\infty} \dfrac{x^{10}}{2^x}$.

SOLUTIONS:

a) At first glance, it appears that the numerator is "more powerful" than the denominator and, if so, $\lim\limits_{x \to \infty} \dfrac{x^{10}}{2^x}$ would equal zero. And for small values of x, this is true. But when $x = 100$, the numerator is approximately 10^{20} while the denominator is approximately 10^{30}. The variable in the exponent is "more powerful" than the variable to any numerical power, so $\lim\limits_{x \to \infty} \dfrac{x^{10}}{2^x} = 0$.

b) Think of plugging in -100 into the expression. You would get $(-100)^{10}(2^{100})$. So it is logical that $\lim\limits_{x \to -\infty} \dfrac{x^{10}}{2^x} = \infty$.

TEST TIP

When you are asked to find horizontal asymptotes to a function f, you are really being asked to find $\lim\limits_{x \to -\infty} f(x)$ and $\lim\limits_{x \to \infty} f(x)$.

EXAMPLE 22: Find horizontal asymptotes for $f(x) = \dfrac{1+e^x}{1-e^x}$.

SOLUTION: $\displaystyle\lim_{x\to\infty}\frac{1+e^x}{1-e^x} = -1$ (This would be a "powers equal" situation.)

$$\lim_{x\to-\infty}\frac{1+e^x}{1-e^x} = \lim_{x\to\infty}\frac{1+\dfrac{1}{e^x}}{1-\dfrac{1}{e^x}} = 1 \quad \left(\lim_{x\to\infty}\frac{1}{e^x} = 0\right)$$

The function has horizontal asymptotes at $y = 1$ and $y = -1$.

EXAMPLE 23: Find a) $\displaystyle\lim_{x\to\infty}\sin x$

b) $\displaystyle\lim_{x\to\infty}\frac{\sin x}{x}$.

SOLUTIONS:
a) Since $\sin x$ oscillates infinitely between 1 and -1, $\displaystyle\lim_{x\to\infty}\sin x$ does not exist.

b) Although $\sin x$ oscillates infinitely between 1 and -1, the denominator will be getting infinitely larger. So this is a "bottom-heavy" expression and $\displaystyle\lim_{x\to\infty}\frac{\sin x}{x} = 0$.

DIDYOU**KNOW?**

The sine and cosine functions are fundamental to the theory of periodic functions as those that describe sound and light waves.

Continuity

Overview: A very loose definition of a **continuous curve** is one that can be drawn without picking up the pencil from the paper. Lines are continuous, parabolas are continuous, and so are sine curves. But for more complex curves, we need a definition that can prove where a function is continuous.

A function $f(x)$ is continuous at $x = c$ if all three conditions hold:

1. $\lim\limits_{x \to c} f(x)$ exists   2. $f(c)$ exists   3. $\lim\limits_{x \to c} f(x) = f(c)$

What this says is that the limit must exist at $x = c$, the function must have a value at $x = c$, and that this limit and value must be the same.

If a function is continuous at all values c in its domain, the function is continuous.

EXAMPLE 24: For each of the following functions, examine their graphs and determine if the function is continuous at the given value of x and, if not, which of the rules of continuity above it fails.

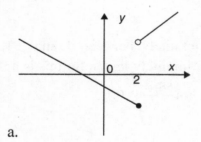

a.

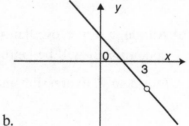

b.

c.

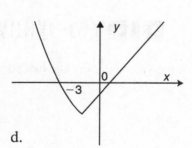

d.

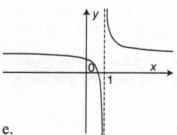

e.

f.

SOLUTIONS:

a. is not continuous at $x = 2$ because $\lim_{x \to 2} f(x)$ does not exist. $\lim_{x \to 2^-} f(x) \neq \lim_{x \to 2^+} f(x)$.

b. is not continuous at $x = 3$ because $f(3)$ does not exist.

c. is not continuous at $x = 0$ because $\lim_{x \to 0} f(x) \neq f(0)$.

d. is continuous at $x = -3$.

e. is not continuous at $x = 1$ because $\lim_{x \to 1} f(x)$ does not exist and $f(1)$ does not exist.

f. While this is not a continuous curve, it is continuous at $x = 5$.

Functions that are examined in algebraic forms have the following characteristics:

- All polynomials are continuous everywhere.

- All functions in the form $\dfrac{f(x)}{g(x)}$ are continuous except where $g(x) = 0$.

- All radicals in the form of $\sqrt[\text{odd root}]{f(x)}$ are continuous everywhere.

- All radicals in the form of $\sqrt[\text{even root}]{f(x)}$ are continuous only when $f(x) \geq 0$.

- Functions in the form of $c^{f(x)}$ where c is a positive constant are continuous for all values where $f(x)$ is continuous.

- Functions involving trigonometry and logarithmic expressions are continuous only within the domain of these expressions.

EXAMPLE 25: Find values of x where the following functions are continuous:

a. $f(x) = x^3 - 2x^2 - 8$

b. $f(x) = \dfrac{x^2 - 25}{x - 5}$

c. $f(x) = \sqrt[3]{4x^2 - 2x - 1}$

d. $f(x) = \sqrt{x^2 + 4x - 32}$

e. $f(x) = \dfrac{\sin x - \ln x}{2^x}$

SOLUTIONS:
a. Continuous everywhere because it is a polynomial.

b. The fact that we can factor the numerator and cancel means we can find $\lim_{x \to 5} f(x)$. But this function is continuous everywhere except at $x = 5$ because $f(5)$ does not exist.

c. Continuous everywhere because it is an odd root.

d. Continuous at all values of x except when $x^2 + 4x - 32 = (x + 8)(x - 4) < 0$. So $f(x)$ is continuous except where $-8 < x < 4$. Or alternatively, f is continuous on $(-\infty, -8]$ and on $[4, \infty)$.

e. Continuous except when $x \le 0$. Or alternatively, continuous on $(0, \infty)$.

TEST TIP

If you are asked values of x where functions are continuous and asked to justify your answers on the AP exams, you may use the characteristics of continuous functions above. But for piecewise functions, you must show work that proves continuity using the 3-part definition of continuity.

EXAMPLE 26: Find the values of x where the following functions are continuous. Justify your answers.

a. $f(x) = \begin{cases} x^2 - x - 4, & x \le 3 \\ 3x - 2, & x > 3 \end{cases}$

b. $f(x) = \begin{cases} \dfrac{2x - 4}{x^2 + 3x - 10}, & x \ne 2 \\ \dfrac{4 - 3x}{1 - 4x}, & x = 2 \end{cases}$

SOLUTIONS:
a. Since both branches of the piecewise function are polynomials, they are continuous. The question is whether the function is continuous at the value where the rule changes.

$$f(x) = \begin{cases} x^2 - x - 4, & x \le 3 \\ 3x - 2, & x > 3 \end{cases}$$

$$\lim_{x \to 3^-} f(x) = (3)^2 - 3 - 4 = 2 \quad \text{and} \quad \lim_{x \to 3^+} f(x) = 3(3) - 2 = 7$$

Since $\lim_{x \to 3} f(x)$ does not exist, $f(x)$ is continuous at all values of x except at $x = 3$.

b. $f(x) = \begin{cases} \dfrac{2(x-2)}{(x+5)(x-2)}, x \neq 2 \\ \dfrac{4-3x}{1-4x}, x = 2 \end{cases} = \begin{cases} \dfrac{2}{x+5}, x \neq 2 \\ \dfrac{4-3x}{1-4x}, x = 2 \end{cases}$

$\displaystyle\lim_{x \to 2} f(x) = \dfrac{2}{2+5} = \dfrac{2}{7}$ $f(2) = \dfrac{4-6}{1-8} = \dfrac{-2}{-7} = \dfrac{2}{7}$

$f(x)$ is continuous at $x = 2$. But it is not continuous at $x = -5$.

EXAMPLE 27:   Find the value(s) of the constant k that makes function f continuous:

$$f(x) = \begin{cases} k^2 - 20x - 1, x < 1 \\ kx + 9, x \geq 1 \end{cases}$$

SOLUTION: $\displaystyle\lim_{x \to 1^-} \left(k^2 - 20x - 1 \right) = \lim_{x \to 1^+} (kx + 9)$

$k^2 - 21 = k + 9$

$k^2 - k - 30 = 0$

$(k - 6)(k + 5) = 0$

$k = 6, k = -5$

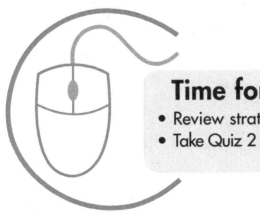

Time for a quiz
- Review strategies in Chapter 2
- Take Quiz 2 at the REA Study Center

(www.rea.com/studycenter)

Derivatives

Average and Instantaneous Rate of Change

Overview: Calculus is all about change and change occurs over time. The change can be quick as in one's position on a highway over a period of a minute. The change can be slow as in the length of one's fingernails over a period of a week. In the latter case, we cannot actually see the growth as it occurs, but we know it does because one week later, your nails, which were short, are now long.

If we were told that a child was growing at the rate of 4 inches/year, we might assume that at any time during the year, his growth rate is 4 inches/year. That is an example of **instantaneous growth**. It is possible though that the child stays at 40 inches tall for 9 months and yet after 12 months, he has grown to 44 inches tall. He grew 4 inches over the year so his **average growth rate** is 4 inches/year yet at any time in those first 9 months, his instantaneous growth rate was 0 inches/year.

To find the average rate of change of y from x_1 to x_2, we use the formula: Avg rate of change $= \dfrac{y_2 - y_1}{x_2 - x_1}$. While the x values are typically units of time, they do not have to be.

EXAMPLE 1: Given $f(x) = x^2 - 2x - \ln x$, find the average rate of change of f from $x = 1$ to $x = e$.

SOLUTION: Avg. rate of change $= \dfrac{e^2 - 2e - 1 - (1 - 2 - 0)}{e - 1} = \dfrac{e^2 - 2e}{e - 1}$

EXAMPLE 2: A particle is moving such that its position on the x-axis is given by $x(t) = 2\cos t + 3\sin t + 2$, where t is measured in seconds and $x(t)$ is measured in meters.

a) Find the average rate of change of its position on the interval $0 \le t \le \dfrac{\pi}{2}$. Specify units.

b) What is the special name of the quantity you found?

SOLUTIONS: Avg. rate of change of position $= \dfrac{x\left(\dfrac{\pi}{2}\right) - x(0)}{\dfrac{\pi}{2} - 0}$

a) $\dfrac{2\cos\dfrac{\pi}{2} + 3\sin\dfrac{\pi}{2} + 2 - (2\cos 0 + 3\sin 0 + 2)}{\dfrac{\pi}{2} - 0}$

$\dfrac{0 + 3 + 2 - 2 - 0 - 2}{\dfrac{\pi}{2}} = \dfrac{2}{\pi}\dfrac{\text{meters}}{\text{sec}}$

b) This is the average velocity for $0 \le t \le \dfrac{\pi}{2}$.

Average rate of change can be important but doesn't necessarily tell us much, especially if the gap in time is large. If you were told that the stock market had an average rate of change of 200 points/month, that doesn't tell you how it fared on a particular day. You can have an average of 80% in calculus over a semester, but if you failed the last two exams, at this particular moment, you are doing failing work.

So we are interested in the instantaneous rate of change or how a quantity is changing at a particular instant. To find it, we need calculus. But for now, we will find an approximation for the instantaneous rate of change.

It is possible to approximate instantaneous rates of change if you are given a table of n values.

x	x_1	x_2	...	x_i	...	x_{n-1}	x_n
y	y_1	y_2	...	y_i	...	y_{n-1}	y_n

To find the approximate instantaneous rate of change at $x = i$, you can use any of these formulas:

$$\text{Instantaneous rate of change} \approx \frac{y_{i+1} - y_i}{x_{i+1} - x_i} \text{ or } \frac{y_i - y_{i-1}}{x_i - x_{i-1}} \text{ or } \frac{y_{i+1} - y_{i-1}}{x_{i+1} - x_{i-1}}$$

Again, while the x values are typically units of time, they do not have to be.

EXAMPLE 3: Values of $f(x)$ are given for selected values of x in the table below. Find

a) the average rate of change of f on the interval [0, 20].
b) the approximate rate of change of f at $x = 12$.
c) the approximate rate of change of f at $x = 20$.

x	0	3	5	8	12	15	20
f(x)	8	10	16	5	-1	-8	0

SOLUTIONS: a) Average rate of change $= \dfrac{f(20) - f(0)}{20 - 0} = \dfrac{0-8}{20} = \dfrac{-2}{5}$

b) Approximate rate of change at time $x = 12$ is either

$$\frac{f(12) - f(8)}{12 - 8} = \frac{-1 - 5}{4} = \frac{-3}{2} \text{ OR}$$

$$\frac{f(15) - f(12)}{15 - 12} = \frac{-8 + 1}{3} = \frac{-7}{3} \text{ OR}$$

$$\frac{f(15) - f(8)}{15 - 8} = \frac{-8 - 5}{7} = \frac{-13}{7}$$

c) Approximate rate of change at time $x = 20$ is

$$\frac{f(20) - f(15)}{20 - 15} = \frac{0 + 8}{5} = \frac{8}{5}$$

EXAMPLE 4: a) Ralph loves pistachio nuts. The total number of nuts he eats is given by a function P of time t. A table of selected values of P for the time interval $0 \le t \le 10$ is shown below, where t is measured in minutes. Use data from the table to approximate the instantaneous rate of change of P at $t = 5$. Show the computation that leads to your answer, explain the meaning of your answer, and specify units of measure.

t (minutes)	0	2	4	6	8	10
$P(t)$ (pistachio nuts)	0	7	18	29	44	62

SOLUTION: Instantaneous rate of change $\approx \dfrac{P(6)-P(4)}{6-4} = \dfrac{29-18}{2} = \dfrac{11}{2}$

At $t = 5$ minutes, Ralph is eating at the rate of approximately 5.5 nuts/minute.

b) Ralph's wife Alice also loves pistachio nuts. Her rate of eating in nuts per minutes is given by the function R of time t. A table of selected values of R for the time interval $0 \le t \le 10$ is below. Use data from the table to approximate the instantaneous rate of change of R at $t = 5$. Show the computation that leads to your answer, explain the meaning of your answer, and specify units of measure.

t (minutes)	0	2	4	6	8	10
$R(t)$ (nuts per minutes)	9	8	7	4	2	3

SOLUTION: Instantaneous rate of change of

$$R \approx \frac{R(6)-R(4)}{6-4} = \frac{4-7}{2} = -\frac{3}{2} \frac{\text{nuts/min}}{\text{min}}.$$

At $t = 5$ min, the rate that Alice eats nuts is decreasing by approximately 1.5 $\dfrac{\text{nuts}}{\text{min}^2}$.

Note the subtle differences between a) and b). In part a), you are given the function that represents the <u>number</u> of nuts that Ralph eats and you are interested in how fast that number is changing. In part b) you are given the <u>rate</u> that Alice eats nuts and you are interested in how fast that rate is changing.

TEST TIP

Beware of the word "approximate." Frequently on the AP test, you are asked to approximate some quantity. In Example 4a, you are asked to approximate the rate of change of P. Whenever you are asked to "approximate," be sure to use an approximate sign (≈) rather than an equal sign. Failure to do so can cost you a point on a free-response question.

The Derivative

Overview: Given a continuous function $f(x)$ over an interval $[a, b]$, we draw a line through the point $(a, f(a))$ and the point $(b, f(b))$. This line is called the **secant line** containing the two points as shown by the dotted line in the figure below. We can find the slope of this line using the familiar precalculus formula: $\text{slope} = m = \dfrac{\text{rise}}{\text{run}} = \dfrac{f(b) - f(a)}{b - a}$.

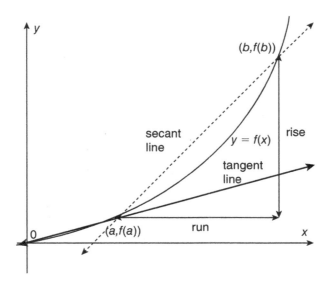

We now concentrate on the **tangent line** at $x = a$. The tangent line intersects (or touches) the curve $f(x)$ in one point rather than two. As the value of b gets closer and closer to a, the secant line starts to look more and more like the tangent line and thus the slope of the secant line starts to approach the slope of the tangent line.

We state this mathematically: If h represents the horizontal distance between the two points a and b, then $h = b - a$ and $b = a + h$. It follows that $m_{sec} = \dfrac{\text{rise}}{\text{run}} = \dfrac{f(a+h) - f(a)}{h}$. As b gets closer and closer to a, we say that $m_{tan} = \lim\limits_{h \to 0} \dfrac{f(a+h) - f(a)}{h}$. Notice that h cannot *equal* zero, so we must express the slope of the tangent line to f at $x = a$ as a limit.

That is the starting point for differential calculus. We now define a **derivative**: a formula for the slope of the tangent line to a function f at any point x. All of differential calculus is based on this definition, so you need to know it.

The relationship that we established early in this chapter can be expressed as:

$$\frac{\text{average rate of change}}{\text{instantaneous rate of change}} = \frac{\text{slope of secant line}}{\text{slope of tangent line (derivative)}}$$

Our usual notation for the derivative can be one of a few. If the function is given in the form of "$f(x) =$", the derivative will be written as $f'(x)$ or f'. If the function is given in the form of "$y =$", the derivative will be written as y or $\dfrac{dy}{dx}$. These notations are interchangeable.

So what you need to know is this:

The informal definition of derivative is a formula for the slope of the tangent line to a curve at a point.

The formal definition of a derivative is $f'(x) = \lim\limits_{h \to 0} \dfrac{f(x+h) - f(x)}{h}$.

A formal definition to find a derivative at a point is $f'(a) = \lim\limits_{x \to a} \dfrac{f(x) - f(a)}{x - a}$.

EXAMPLE 5: Using the formal definition, find the derivative of $f(x) = x^2 + 7x - 2$, and use it to find the slope of the tangent line to f at $x = -2$.

SOLUTION:

$$f'(x) = \lim_{h \to 0} \frac{f(x+h) - f(x)}{h} = \lim_{h \to 0} \frac{(x+h)^2 + 7(x+h) - 2 - (x^2 + 7x - 2)}{h}$$

Using our technique of finding limits, realize that plugging in $h = 0$ will give a zero in the denominator. So we have to hope we can do some factoring and then canceling. In this type of problem, we always can.

$$f'(x) = \lim_{h \to 0} \frac{x^2 + 2xh + h^2 + 7x + 7h - 2 - x^2 - 7x + 2}{h}$$

$$f'(x) = \lim_{h \to 0} \frac{2xh + h^2 + 7h}{h}$$

$$f'(x) = \lim_{h \to 0} \frac{h(2x + h + 7)}{h}$$

$$f'(x) = \lim_{h \to 0} (2x + h + 7)$$

$$f'(x) = 2x + 7$$

slope of tangent line at $x = -2 = f'(-2) = 2(-2) + 7 = 3$

EXAMPLE 6:   Using the secondary definition, find $f'(5)$ if $f(x) = \dfrac{3}{x-2}$.

SOLUTION:

$$f'(5) = \lim_{x \to 5} \frac{f(x) - f(5)}{x - 5}$$

$$f'(5) = \lim_{x \to 5} \left(\frac{\dfrac{3}{x-2} - 1}{x - 5} \right) \left(\frac{x-2}{x-2} \right)$$

$$f'(5) = \lim_{x \to 5} \frac{3 - x + 2}{(x-5)(x-2)}$$

$$f'(5) = \lim_{x \to 5} \frac{5 - x}{(x-5)(x-2)}$$

$$f'(5) = \lim_{x \to 5} \frac{-1(x-5)}{(x-5)(x-2)}$$

$$f'(5) = \lim_{x \to 5} \frac{-1}{x-2} = -\frac{1}{3}$$

TEST TIP

Very rarely will you ever be asked to take the derivative of a function by using the definition. There are better ways, covered in the following several sections. However, you must understand the definition so that you know that $\lim\limits_{\Delta x \to 0} \dfrac{\cos(1+\Delta x) - \cos(1)}{\Delta x}$ is equivalent to the slope of the tangent line to $y = \cos x$ at $x = 1$. Doing this by the definition is cumbersome. Using the derivatives rules below makes the problem quite simple. Also note that the usual h in the definition is replaced by Δx which means change in x. It makes no difference what variable is used to represent this change.

Derivatives of Algebraic Functions

Overview: Differential calculus requires you to be able to take derivatives quickly and efficiently. For algebraic functions, there are four rules that you must know backwards and forwards.

$$\text{Constant Rule: } \frac{d}{dx}(C) = 0$$

$$\text{Power Rule: } \frac{d}{dx}(x^n) = nx^{n-1}$$

$$\text{Product Rule: } \frac{d}{dx}[f(x) \cdot g(x)] = f(x) \cdot g'(x) + g(x) \cdot f'(x)$$

$$\text{Quotient Rule: } \frac{d}{dx}\left[\frac{f(x)}{g(x)}\right] = \frac{g(x) \cdot f'(x) - f(x) \cdot g'(x)}{[g(x)]^2}$$

The process of taking derivatives is called differentiation.

EXAMPLE 7: If $f(x) = (x^2 + 6x + 10)(2x^2 - 7x - 1)$, find $f'(-2)$.

SOLUTION: Use the product rule as it is too time-consuming to expand the expression.

$$f'(x) = (x^2 + 6x + 10)(4x - 7) + (2x^2 - 7x - 1)(2x + 6)\,(-2).$$

If you were asked to find $f'(x)$ in a simplified form you would need to expand this expression. But you only want $f'(-2)$ so simply evaluate the expression at $x = -2$.

$$f'(-2) = (2)(-15) + (21)(2) = 12$$

EXAMPLE 8: Given $y = \dfrac{7x^2 - 9x - 3}{x}$, find $\dfrac{dy}{dx}$.

SOLUTIONS: There are three ways to handle this problem:

a) The quotient rule: $\dfrac{dy}{dx} = \dfrac{x(14x - 9) - (7x^2 - 9x - 3)1}{x^2} = \dfrac{7x^2 + 3}{x^2}$

b) Splitting it: $y = \dfrac{7x^2}{x} - \dfrac{9x}{x} - \dfrac{3}{x} = 7x - 9 - \dfrac{3}{x}$

To take the derivative of $\dfrac{3}{x}$, you could either use the quotient rule or write it as $3x^{-1}$. In any case, the final answer is $\dfrac{dy}{dx} = 7 + \dfrac{3}{x^2}$. While not in the same form as the quotient rule method above, the two are equivalent by using a common denominator.

c) Using the product rule:

$$y = \frac{7x^2 - 9x - 3}{x} = (7x^2 - 9x - 3)(x^{-1})$$

$$\frac{dy}{dx} = (7x^2 - 9x - 3)(-x^{-2}) + (x^{-1})(14x - 9)$$

$$\frac{dy}{dx} = \frac{-(7x^2 - 9x - 3)}{x^2} + \frac{14x - 9}{x}$$

This answer is equivalent to the others by expressing with a common denominator. But it takes a few steps.

TEST TIP

Which way do you handle derivatives involving quotients? The quotient rule is difficult for some students and many will do anything to avoid using it. But using the product rule ends up involving negative exponents, which are rarely in AP exam multiple-choice answers. If a problem can be split into individual fractions which have their derivatives easily taken, do so. But sometimes you have to live with the quotient rule. So learn it and remember that if your answer doesn't match one of the given multiple-choice answers, a little algebraic cleanup might be necessary.

EXAMPLE 9: If $f(x) = \dfrac{ax+b}{ax-b}$ where a and b are constants, $a \neq 0, b \neq 0$, for what values of a and b is $f'(x) > 0$?

SOLUTION: $f'(x) = \dfrac{(ax-b)a-(ax+b)a}{(ax-b)^2} = \dfrac{-2ab}{(ax-b)^2}$

Since the denominator is always positive, $f'(x) > 0$ when a and b have opposite signs.

TEST TIP

While the TI-84 calculator cannot actually find a derivative, it can approximate the derivative of a function *at a point*. Some teachers do not actually teach this process, but it is important to be able to calculate derivatives using technology, especially when taking the derivative is too complex. The AP exam requires students to be skilled in using a graphing calculator. There are two ways to approximate the derivative: use the NDERIV function from the MATH menu or use the dy/dx option from the GRAPH menu.

EXAMPLE 10: Given $f(x) = 2x^2 + 10x - \dfrac{1}{x}$, find $f'(-3)$.

SOLUTION: Algebraically,

$$f'(x) = 4x + 10 + \frac{1}{x^2} \text{ and } f'(-3) = -12 + 10 + \frac{1}{9} = \frac{-17}{9}.$$

Graphically:

Plot1 Plot2 Plot3
\Y₁=2X²+10X-1/X
\Y₂=
\Y₃=
\Y₄=
\Y₅=
\Y₆=
\Y₇=

```
nDeriv(Y1, X ,-3)
          -1.888888877
-17/9
          -1.888888889
```

dy/dx=-1.888889

TEST TIP

There are occasionally times when the calculator's approximation of the derivative is completely wrong and, later in this chapter, you will learn when to recognize those situations. Remember though that in the vast majority of occurrences when you use the calculator to approximate a derivative at a point, the value it displays might be slightly off. Still, if you set the calculator to three decimal places, the AP standard, the approximation of the derivative will be completely accurate.

Since the AP test writers know that students have the ability to find derivatives at points using a calculator, they devise problems in which having the use of the calculator does students no good. These problems involve the use of tables and these types of problems invariably show up in the multiple-choice section. Example 11 is a case in point.

EXAMPLE 11: The table below gives values of the differentiable function f and g at $x = 4$. If $h(x) = \dfrac{g(x) - f(x)}{2f(x)}$, then $h'(4) = ?$

x	$f(x)$	$g(x)$	$f'(x)$	$g'(x)$
4	-2	e	-1	3

SOLUTION: You can either treat the function h as a quotient or split it into two parts to evaluate its derivative. Splitting is probably easier.

$$h(x) = \frac{1}{2}\frac{g(x)}{f(x)} - \frac{1}{2}\frac{f(x)}{f(x)} = \frac{1}{2}\frac{g(x)}{f(x)} - \frac{1}{2}$$

$$h'(x) = \frac{1}{2}\left[\frac{f(x)g'(x) - g(x)f'(x)}{[f(x)]^2}\right]$$

$$h'(4) = \frac{1}{2}\left[\frac{f(4)g'(4) - g(4)f'(4)}{[f(4)]^2}\right] = \frac{1}{2}\left[\frac{-2(3) - e(-1)}{(-2)^2}\right]$$

$$h'(4) = \frac{e - 6}{8}$$

You can and will have to take 2nd derivatives (derivatives of derivatives) and beyond. In the next chapter we will learn their meaning and application. The following table shows proper notation:

derivative	notation
1st	$f'(x)$ or $\dfrac{dy}{dx}$
2nd	$f''(x)$ or $\dfrac{d^2y}{dx^2}$
3rd	$f'''(x)$ or $\dfrac{d^3y}{dx^3}$
4th	$f^{(4)}(x)$ or $\dfrac{d^4y}{dx^4}$
nth	$f^{(n)}(x)$ or $\dfrac{d^ny}{dx^n}$

EXAMPLE 12: For what values of x is $f'(x) = f''(x)$ for $f(x) = \dfrac{x^2+x-1}{x}$?

SOLUTION:

$$f(x) = \frac{x^2+x-1}{x} = x+1-\frac{1}{x}$$

$$f'(x) = 1+\frac{1}{x^2} \text{ and } f''(x) = \frac{-2}{x^3}$$

$$1+\frac{1}{x^2} = \frac{-2}{x^3}$$

$$x^3+x = -2 \text{ or } x^3+x+2 = 0$$

Since no calculator is allowed, the solution must be one that can be "eyeballed" if it is a free-response problem. If it is a multiple-choice problem, each of the choices can be plugged into this equation.

The answer is $x = -1$.

The Chain Rule

Overview: There is no more important rule relating to differentiation than the chain rule. Rather than trying to identify the form of a function (such as product or quotient), and then applying the derivative rule, the chain rule is a technique that oversees all the rules of differentiation. The chain rule is always in effect!

The chain rule can be expressed several ways and can be confusing to students. Like trying to explain how to ride a bike, sometimes it is easier just to try it.

$$\text{If } y = f(u) \text{ and } u = g(x), \frac{dy}{dx} = \frac{dy}{du} \cdot \frac{du}{dx} \quad \text{or} \quad \frac{d}{dx} f(g(x)) = f'(g(x)) \cdot g'(x)$$

EXAMPLE 13: If $f(x) = \sqrt{4x+1}$, find $f'(x)$ and $f''(x)$.

SOLUTION: $f'(x) = (4x+1)^{1/2}$

The differentiation technique requires you to treat the $(4x + 1)^{1/2}$ as an entity and apply the power rule to it.

$$f'(x) = \frac{1}{2}(4x+1)^{-1/2}(4) = \frac{2}{\sqrt{4x+1}}$$

$$f''(x) = 2\left(\frac{-1}{2}\right)(4x+1)^{-3/2}(4) = \frac{-4}{(4x+1)^{3/2}}$$

EXAMPLE 14: If $y = \sqrt{\dfrac{2x-2}{3x+7}}$, find the slope of the line normal to the graph of y at $x = 3$.

SOLUTION: $y' = \dfrac{1}{2}\left(\dfrac{2x-2}{3x+7}\right)^{-1/2}\left[\dfrac{(3x+7)2-(2x-2)3}{(3x+7)^2}\right]$

The expression in brackets is the derivative of $\left(\dfrac{2x-2}{3x+7}\right)$ using the quotient rule.

$y' = \dfrac{1}{2}\left(\dfrac{3x+7}{2x-2}\right)^{1/2}\left[\dfrac{20}{(3x+7)^2}\right]$

$y'(3) = \sqrt{\dfrac{16}{4}}\left(\dfrac{10}{16^2}\right) = \dfrac{5}{64}$

Slope of normal line at $x = 3$ is $\dfrac{-64}{5}$.

EXAMPLE 15: The table below gives values of the functions f and g and their derivatives at selected values of x with a being a constant. If the slope of the tangent line to $f(g(x))$ at $x = 1$ is 5, find the value of a.

x	$f(x)$	$f'(x)$	$g(x)$	$g'(x)$
1	8	3	2	$\dfrac{1}{a}$
2	−2	−4	2	6

SOLUTION: $\dfrac{d}{dx}f(g(x)) = f'(g(x))\cdot g'(x)$

$f'(g(1))\cdot g'(1) = f'(2)\cdot\dfrac{1}{a} = 5$

$\dfrac{-4}{a} = 5$

$a = \dfrac{-4}{5}$

DIDYOUKNOW?

There are many different notations for derivatives. The earliest, $\frac{dy}{dx}$, was introduced by Gottfried Wilhelm Leibniz (1646–1716). Joseph-Louis Lagrange (1736–1813) introduced the prime notation such as f' and f''. However there are several more obscure notations that still survive. Leonhard Euler (1707-1873) used the notation $D_x y$ as the first derivative and $D^2_x y$ as the second derivative. Sir Isaac Newton (1642–1727) used dot notation, representing the first two derivatives as $\dot{y}$ and $\ddot{y}$.

Tangent Line Approximations

Overview: Now that we can calculate the slope of the tangent line by computing the derivative of f and evaluating it at some value of x, we take the next step and actually find the equation of the tangent line.

Recall that if we have a graph of $y = f(x)$ passing through a point (x_1, y_1), the equation of the tangent line using the **point-slope formula** is: $y - y_1 = m(x - x_1)$. Whenever you are asked for an equation of a line, that is the equation you will need. Typically students associate equations of lines with the formula $y = mx + b$, the slope-intercept form. While correct, the point-slope formula is used almost exclusively in calculus. In the figure below, the straight line is tangent to the curve at $x = 1$.

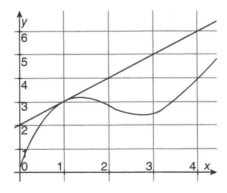

Realize that when you use the point-slope formula $y - y_1 = m(x - x_1)$, the slope m is the derivative of the function f at point (x_1, y_1). This allows you to calculate the equation of the tangent line to a graph easily.

Finally, you can plug in some value of c close to x_1 to estimate $f(c)$. This is called the **tangent-line approximation** and the closer c is to x_1, the better the approximation will be. This uses the property of **local linearity**, which says the more you zoom-in on a curved graph, the more the graph looks like a straight line. In the graph on the previous page, to approximate the value of the function f at $x = 1.1$, we could evaluate the value of the line at $x = 1$. Since 1.1 is close to 1, the approximation will be fairly accurate. There are techniques that are covered in Chapter 6 to determine whether it is an under or over-approximation (as it is in this case) without actually seeing the graphs of the function and the line.

EXAMPLE 16:   The graph below is $f(x) = 12 - x - x^2$. Find the equation of the tangent line to $f(x)$ at $x = 2$ and use it to estimate $f(1.9)$.

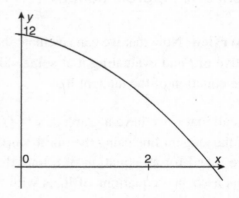

SOLUTION: Point: $(2, 12 - 2 - 4) = (2, 6)$

Slope: $\dfrac{dy}{dx} = -1 - 2x$ so $m = \dfrac{dy}{dx}\Big|_{x=2} = -1 - 4 = -5$

Equation: $y - 6 = -5(x - 2)$ or $y = 16 - 5x$

This can be verified graphically: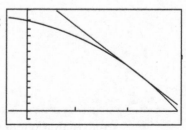

So $f(1.9) \approx 16 - 5(1.9) = 6.5$.

The TI−84 has the capability of finding the equation of the tangent line to a function at a point. It is found in the DRAW menu and must be accessed when viewing the graph.

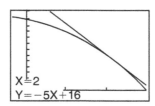

```
DRAW POINTS STO
1: Clr Draw
2: Line(
3: Horizontal
4: Vertical
5: Tangent(
6: DrawF
7↓ Shade(
```

X=2
Y=−5X+16

TEST TIP

While helpful, you may never use the TANGENT feature as a justification on an AP exam free-response question. The only features of the calculator (other than basic calculations) you may use without justification are 1) evaluating functions at x-values (CALC-1:Value), 2) finding zeros of functions or intersection of two functions (CALC-2:Zero or CALC-5:Intersect), 3) approximating the derivative at a point using NDERIV, and 4) approximating the definite integral between two values (covered in Chapter 7) using FNINT.

EXAMPLE 17: Suppose that function f has a continuous first derivative for all x and that $f(0) = 5$ and $f'(0) = 7$. Let g be a function given by $g(x) = (2x-3)^2 f(x)$ for all x. Write an equation of the tangent line to the graph of g at $x = 0$ and use it to approximate $f(0.25)$.

SOLUTION: It looks confusing, but as soon as you see the words "equation of tangent line," you think: $y - y_1 = m(x - x_1)$.

You need the point: $g(0) = (0-3)^2 f(0) = 9(5) = 45$

You need the slope: Use the product rule and chain rule.

$g'(x) = (2x-3)^2 f'(x) + f(x)(2)(2x-3)2$

$g'(0) = (-3)^2 f'(0) + f(0)(2)(-3)2$

$g'(0) = 9(7) + 5(2)(-3)2 = 63 - 60 = 3$

Tangent line: $y - 45 = 3(x-0)$ or $y = 3x + 45$

Approximation: $f(0.25) \approx 3(0.25) + 45 = 45.75$

EXAMPLE 18: The function f is continuous and $f(-4) = 6$ and $f'(-4) = 3$. For what value of k is the approximation of $f(k)$ using the tangent line of f at $x = k$ equal to k^2?

SOLUTION: Tangent line: $y - 6 = 3(x + 4) \Rightarrow y = 3x + 18$

$f(k) = 3k + 18 = k^2$

$k^2 - 3k - 18 = 0$

$(k + 3)(k - 6) = 0$

$k = -3, \ k = 6$

Derivatives of Trigonometric Functions

Overview: You must know the derivatives of the trigonometric functions, especially for sine, cosine, and tangent as given below. The co-functions (csc, sec, cot) can be derived using the three basic functions and the quotient rule, but they are best memorized as at least one of them is sure to show up on the AP exam. It is also very handy to keep the trig identities $\tan x = \dfrac{\sin x}{\cos x}$ and $\sin^2 x + \cos^2 x = 1$ in mind as it makes some complicated derivatives easier to compute.

$$\frac{d}{dx}(\sin x) = \cos x \qquad \frac{d}{dx}(\cos x) = -\sin x \qquad \frac{d}{dx}(\tan x) = \sec^2 x$$

$$\frac{d}{dx}(\csc x) = -\csc x \cot x \qquad \frac{d}{dx}(\sec x) = \sec x \tan x \qquad \frac{d}{dx}(\cot x) = -\csc^2 x$$

EXAMPLE 19: If $f(x) = \sin^3 5x$, find $f'(x)$.

SOLUTION: The key to problems like this is rewriting the problem using parentheses and brackets: $f(x) = [\sin(5x)]^3$

To take the derivative, realize that this problem utilizes first the power rule and then trig derivatives with the chain rule in play: $f'(x) = 3[\sin(5x)]^2 [\cos(5x)]5$

Final answer: $f'(x) = 15[\sin(5x)]^2 [\cos(5x)]$ or $f'(x) = 15\sin^2 5x \cos 5x$.

EXAMPLE 20: If $f(x) = \sqrt{\cos(2x)}$, find $f'\left(\dfrac{\pi}{6}\right)$.

SOLUTION:

$$f(x) = \left[\cos(2x)\right]^{1/2}$$

$$f'(x) = \frac{1}{2}\left[\cos(2x)\right]^{-1/2}\left[-\sin(2x)\right]2$$

$$f'(x) = \frac{-\sin(2x)}{\sqrt{\cos(2x)}}$$

$$f'\left(\frac{\pi}{6}\right) = \frac{-\sin(\pi/3)}{\sqrt{\cos(\pi/3)}} = \frac{-\sqrt{3}/2}{\sqrt{1/2}} = \frac{-\sqrt{3}}{2} \cdot \frac{\sqrt{2}}{1} = \frac{-\sqrt{6}}{2}$$

EXAMPLE 21: What is the equation of the line tangent to the graph of

$$y = 4\sin x \cos x + 3\sin x \quad \text{at } x = \frac{3\pi}{2}?$$

SOLUTION:

$$y' = 4[\sin x(-\sin x) + \cos x(\cos x)] + 3\cos x$$
$$= 4(\cos^2 x - \sin^2 x) + 3\cos x$$

$$y'\left(\frac{3\pi}{2}\right) = 4(0-1) + 3(0) = -4$$

$$y\left(\frac{3\pi}{2}\right) = 4(-1)(0) + 3(-1) = -3$$

Tangent line: $y + 3 = -4\left(x - \dfrac{3\pi}{2}\right)$

EXAMPLE 22: (Calculator Active) If $f(x) = \tan x - x^2$ and $g(x) = \sec x$, find all values of x on $\left[0, \dfrac{\pi}{4}\right]$ for which $f'(x) = g'(x)$.

SOLUTION:

$$f'(x) = \sec^2 x - 2x, \ g'(x) = \sec x \tan x$$
$$\sec^2 x - 2x = \sec x \tan x \text{ at } x = 0.368$$

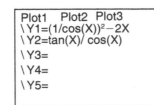
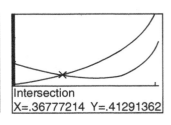

TEST TIP

Be sure you know how to input the cosecant, secant, and cotangent functions into your calculator using reciprocals. While a problem like Example 22 is easy to do once you take the derivatives (because the calculator does the work), improper input of the functions will lead to incorrect answers. On multiple-choice problems, if your answer is not one of the choices, you know you made a mistake. Be very careful when you input your equations into the calculator or check your calculator afterwards. Preferably both!

Derivatives of Exponential, Logarithmic, and Inverse Trig Functions

Overview: Depending on the calculus course you took, you might have learned the derivatives of these non-polynomial expressions when you first learned about derivatives or they were covered much later in the course. No matter when you were taught, they must be part of your knowledge, especially the derivative of the natural log function ($y = \ln x$) and the exponential function ($y = e^x$). The inverse trig problems show up at least once on an AP exam. Although they can be developed using implicit differentiation (covered later in this chapter), it is best to memorize them.

$$\frac{d}{dx}(\ln x) = \frac{1}{x} \qquad \frac{d}{dx}(e^x) = e^x \qquad \frac{d}{dx}(a^x) = a^x \ln a$$

$$\frac{d}{dx}(\sin^{-1} x) = \frac{1}{\sqrt{1-x^2}} \qquad \frac{d}{dx}(\cos^{-1} x) = \frac{-1}{\sqrt{1-x^2}} \qquad \frac{d}{dx}(\tan^{-1} x) = \frac{1}{1+x^2}$$

When working with the natural log function (ln), students should always remember the following three rules. Using them can simplify expressions before taking derivatives.

$$\ln(ab) = \ln a + \ln b \qquad \ln\left(\frac{a}{b}\right) = \ln a - \ln b \qquad \ln(a^b) = b \ln a$$

DIDYOU**KNOW**?

"Who has not been amazed to learn that the function $y = e^x$, like a phoenix rising from its own ashes, is its own derivative?"—Francois le Lionnais

EXAMPLE 23:   If $f(x) = \ln\left(x^2 - \dfrac{1}{x^2}\right)$, show that $f'(x) = \dfrac{2x^4 + 2}{x^5 - x}$.

SOLUTION: Since this problem's format is the natural log of an expression, we can use the natural log rule above incorporating the chain rule as well.

$$f'(x) = \left(\frac{1}{x^2 - \dfrac{1}{x^2}}\right)\left(2x + \frac{2}{x^3}\right)$$

Don't panic with an expression like this. Simply multiply every term by the least common denominator (LCD) which is x^3 and all the complex fractions will magically disappear.

$$f'(x) = \left(\frac{1}{x^2 - \dfrac{1}{x^2}}\right)\left(\frac{2x + \dfrac{2}{x^3}}{1}\right)\left(\frac{x^3}{x^3}\right) = \frac{2x^4 + 2}{x^5 - x}$$

TEST **TIP**

Sometimes on the AP exam free-response questions, you are asked to show that the derivative of a function is some expression rather than actually finding the derivative of the expression. This is done for two reasons: First, for students who have symbolic graphing calculators like the TI-89, any advantage for them is eliminated. Second, if there is a secondary part of the problem that utilizes the derivative, students will be able to tackle this part even though they may not be able to actually find the derivative.

EXAMPLE 24: If $y = \ln\sqrt{\dfrac{\sin x + 1}{x^2 + 1}}$, find $y'(\pi)$.

SOLUTION:
$$y = \ln\sqrt{\frac{\sin x + 1}{x^2 + 1}}$$

$$y = \ln\left(\frac{\sin x + 1}{x^2 + 1}\right)^{1/2} = \frac{1}{2}\ln\left(\frac{\sin x + 1}{x^2 + 1}\right) = \frac{1}{2}\left[\ln(\sin x + 1) - \ln(x^2 + 1)\right]$$

Notice how using the ln rules simplifies the expression so that taking the derivative is easier.

$$y' = \frac{1}{2}\left[\frac{\cos x}{\sin x + 1} - \frac{2x}{x^2 + 1}\right]$$

$$y'(\pi) = \frac{1}{2}\left[-1 - \frac{2\pi}{\pi^2 + 1}\right]$$

EXAMPLE 25: If $f(x) = \dfrac{\ln x}{e^{2x-1}}$, use the tangent line approximation to f at $x = 1$ to approximate $f(1.5)$.

SOLUTION:
$$f'(x) = \frac{\left(e^{2x-1}\right)\dfrac{1}{x} - (\ln x)(2)\left(e^{2x-1}\right)}{\left(e^{2x-1}\right)^2}$$

$$f'(1) = \frac{e}{e^2} = \frac{1}{e} \qquad f(1) = 0$$

Tangent line: $y = \dfrac{1}{e}(x - 1)$

$$f(1.5) \approx \frac{1}{e}(1.5 - 1) = \frac{1}{2e}$$

TEST TIP

Solving AP questions rarely requires a lot of steps. The testers know your time is limited. Problems such as Example 25 tend to have relatively simple answers with cancellation or zeros occurring along the way. If you see you are getting bogged down in a solution, it is best to take a step back and see if you have made a careless error or if the problem can be attacked in a different way.

EXAMPLE 26: Let f be a function that is differentiable for all values of x. Values of the function and its derivative $f'(x)$ are given in the table below. Let $g(x) = e^{f(x^2)}$ and $g'(-1) = 16$. Find a.

x	$f(x)$	$f'(x)$
1	a	-4
-1	3	0

SOLUTION:

$$g'(x) = e^{f(x^2)}\left[f'(x^2)\right](2x)$$

$$g'(-1) = e^{f(1)}\left[f'(1)\right](-2) = 16$$

$$16 = e^a(-4)(-2)$$

$$e^a = 2 \text{ so } a = \ln 2$$

TEST TIP

Example 26 is a very typical AP problem. It tests students' knowledge of taking derivatives with e, and liberal use of the chain rule. Still, the problem can be done in several steps. Even if it appears on the calculator-active section, calculators are useless. Note that the information given at $x = -1$ has no bearing on the problem.

EXAMPLE 27:   (Calculator Active) Let $y = \sin^{-1}(x^2 - 1)$. For what values of x is $\dfrac{dy}{dx} > 1$?

SOLUTION: $\dfrac{dy}{dx} = \dfrac{2x}{\sqrt{1 - (x^2 - 1)^2}} > 1$

$x > 0.632$ (but see below)

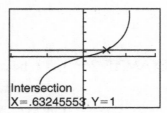

This is verified in the graph below of $y = \sin^{-1}(x^2 - 1)$. Note how the slope is positive when x is positive and it gets steeper as x gets larger. However, the largest value of the sine function is 1 so the largest possible value for x is $\sqrt{2}$. The true solution to this problem is $0.632 < x \leq \sqrt{2}$.

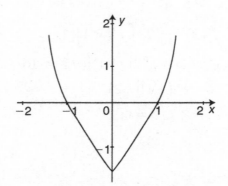

TEST TIP

Although you are being tested on calculus and the testers try hard to assess just those abilities, you cannot throw out algebraic techniques and concepts that are the building blocks of calculus. Concepts like domain that appear in Example 27, are fair game on the AP exam. If you are weak on these concepts, go back and review them. If in a multiple-choice problem, you see two answers: (A) $x > 0.632$ and (B) $0.632 < x \leq \sqrt{2}$, even though you get answer (A), take the time to at least consider answer (B) and avoid choosing the obvious, but incorrect answer, choice (A).

EXAMPLE 28: Find $\lim\limits_{\Delta x \to 0} \dfrac{\cos^2(x + \Delta x) - \cos^2 x}{\Delta x}$.

SOLUTION: This looks like a difficult limit problem. Plugging in $\Delta x = 0$ gives a zero in the denominator and expanding the numerator is quite difficult. Yet this problem can be done in several seconds.

It all has to do with recognition. You should know that

$f'(x) = \lim\limits_{\Delta x \to 0} \dfrac{f(x + \Delta x) - f(x)}{\Delta x}$ and this is the form of the given expression. So you are merely being asked to find $f'(x)$ when $f(x) = \cos^2 x$.

$f(x) = [\cos x]^2$

$f'(x) = 2 \cos x(- \sin x) = - 2\sin x \cos x$

TEST TIP

A problem similar to Example 28 is almost guaranteed to be on the AP exam. While you are rarely asked to take the derivative of functions by definition (using limit techniques), you must recognize a problem expressed in this way for what it is—a derivative problem in disguise!

Implicit Differentiation

Overview: When a function is written in the form "$y =$" or "$f(x) =$", it is said to be in **explicit form**. To differentiate, take the derivative of both sides of the equation. The left side will be $\dfrac{dy}{dx}$ or $f'(x)$ and the right side will be a function of x. But frequently, expressions are not written in the form of "$y =$" because it is too difficult to solve for y as y appears on both sides of the equation. Expressions (that may or may not be functions) written like this are said to be written in **implicit form**.

To find $\dfrac{dy}{dx}$ in a function written implicitly, take the derivative of every term in the given expression, remembering that the chain rule is used and the derivative of y is not

1, but $\dfrac{dy}{dx}$. If you are asked to find $\dfrac{dy}{dx}$ at a given point, rather than solve for $\dfrac{dy}{dx}$, it is best to then plug in the point, eliminate all variables, and then solve for $\dfrac{dy}{dx}$.

EXAMPLE 29: Find the equation of the tangent line to $y^2 - 3x^2 = 5xy + x - 2y$ at $(3,-2)$.

SOLUTION: First realize that solving for y is too difficult.

$$2y\frac{dy}{dx} - 6x = 5\left(x\frac{dy}{dx} + y\right) + 1 - 2\frac{dy}{dx}$$

Evaluate at $(3, -2)$:

$$-4\frac{dy}{dx} - 18 = 5\left(3\frac{dy}{dx} - 2\right) + 1 - 2\frac{dy}{dx}$$

$$-4\frac{dy}{dx} - 18 = 15\frac{dy}{dx} - 10 + 1 - 2\frac{dy}{dx}$$

$$-17\frac{dy}{dx} = 9$$

$$\frac{dy}{dx} = \frac{-9}{17}$$

Tangent line: $y + 2 = \dfrac{-9}{17}(x - 3)$

EXAMPLE 30: The graph of the function $\sin(x + y) = y$ passes through the point $(\pi, 0)$. There is a number k such that the point $(3, k)$ is on the curve. Use the tangent line at $x = \pi$ to approximate k.

SOLUTION:

$$\cos(x + y)\left(1 + \frac{dy}{dx}\right) = \frac{dy}{dx}$$

$$\cos(\pi + 0)\left(1 + \frac{dy}{dx}\right) = \frac{dy}{dx}$$

$$-1 - \frac{dy}{dx} = \frac{dy}{dx}$$

$$\frac{dy}{dx} = \frac{-1}{2}$$

Tangent line: $y = \dfrac{-1}{2}(x - \pi)$

$$k = \frac{-1}{2}(3 - \pi) \approx .071$$

EXAMPLE 31: The graphs of $y + \ln y = x$ and $y + e^y = x$ are shown in the figure below. Show that the tangent lines to both curves at $x = 1$ are parallel.

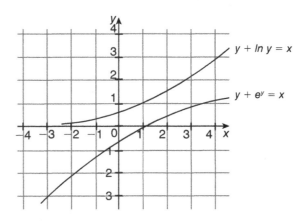

SOLUTION: $y + \ln y = x$ passes through $(1,1)$ $\quad y + e^y = x$ passes through $(1,0)$

$$\frac{dy}{dx} + \frac{1}{y}\frac{dy}{dx} = 1 \qquad\qquad \frac{dy}{dx} + e^y\frac{dy}{dx} = 1$$

$$\frac{dy}{dx}\left(1 + \frac{1}{y}\right) = 1 \qquad\qquad \frac{dy}{dx}\left(1 + e^y\right) = 1$$

$$\frac{dy}{dx} = \frac{1}{1 + \dfrac{1}{y}} = \frac{y}{y+1} \qquad\qquad \frac{dy}{dx} = \frac{1}{1 + e^y}$$

$$\left.\frac{dy}{dx}\right|_{(1,1)} = \frac{1}{2} \qquad\qquad \left.\frac{dy}{dx}\right|_{(1,0)} = \frac{1}{2}$$

Since both lines have the same slope, they are parallel.

EXAMPLE 32: For $x^2 + y^2 = 81$, show that $\dfrac{d^2y}{dx^2} = \dfrac{-81}{y^3}$

SOLUTION:

$$2x + 2y\frac{dy}{dx} = 0$$

$$\frac{dy}{dx} = \frac{-x}{y}$$

$$\frac{d^2y}{dx^2} = \frac{-y + x\dfrac{dy}{dx}}{y^2} = \frac{-y + x\left(\dfrac{-x}{y}\right)}{y^2}$$

$$\frac{d^2y}{dx^2} = \left(\frac{-y - \dfrac{x^2}{y}}{y^2}\right)\left(\frac{y}{y}\right) = \frac{-y^2 - x^2}{y^3}$$

$$\frac{d^2y}{dx^2} = \frac{-\left(x^2 + y^2\right)}{y^3}$$

$$\frac{d^2y}{dx^2} = \frac{-81}{y^3}$$

Derivatives of Inverse Functions

Overview: Inverses are confusing. Most students think that the inverse of 5 is -5, which is not true. Others confuse the inverse of 5 with $\dfrac{1}{5}$, which is the reciprocal of 5. *There is no such thing as the inverse of 5.* We take inverses of functions, not numbers. The inverse of a function f is another function f^{-1} that "undoes" what f does. So $f^{-1}(f(x)) = x$. For instance, the inverse of adding is subtracting. Start with any number x, add 8,247 and then subtract 8,247, you know you are back to x without having to do any calculations.

To find the **inverse of a function**, you replace x with y and y with x. The inverse to the function $y = 3x + 7$ is $x = 3y + 7$ or $y = \dfrac{x - 7}{3}$. The inverse of a function may not be a function. In this section, you are concerned with finding the derivative of the inverse to a function: $\dfrac{d}{dx}[f(x)]^{-1}$.

You find this derivative by using implicit differentiation:

$$x = f(y) \quad \text{so} \quad 1 = f'(y)\frac{dy}{dx} \quad \text{and} \quad \frac{dy}{dx} = \frac{1}{f'(y)}.$$

In a problem, you can use implicit differentiation, or just memorize the formula: $\dfrac{dy}{dx} = \dfrac{1}{f'(y)}$.

EXAMPLE 33: If $f(x) = x^3 + x + 1$, and $g(x) = f^{-1}(x)$, what is the value of $g'(3)$?

SOLUTION: Step 1: First find the inverse to f: $x = y^3 + y + 1$

Step 2: Plug in 3 for x and solve for y. If calculators are not allowed, the solution will be easy to find by graphing. If not, use simple trial and error: $3 = y^3 + y + 1$ and $y = 1$.

Step 3: Either use the formula or implicitly differentiate your step 1 equation: $1 = 3y^2 \dfrac{dy}{dx} + \dfrac{dy}{dx}$ and $\dfrac{dy}{dx} = \dfrac{1}{3y^2 + 1}$.

Step 4: Plug in your value of y from step 2: $\dfrac{dy}{dx} = \dfrac{1}{3(1)^2 + 1} = \dfrac{1}{4}$.

EXAMPLE 34: In the chart below, two values of x are given along with the values of the differentiable function $f(x)$ as well as $f'(x)$. If f^{-1} is the inverse function of f, a) find the derivative of f^{-1} at $x = 3$. That is, find $\left[f^{-1}(3) \right]'$ and b) find the equation of the line tangent to the graph of $y = f^{-1}(x)$ at $x = 3$.

x	−3	3
$f(x)$	3	8
$f'(x)$	−2	4

SOLUTIONS: a) $(-3, 3)$ is on f so $(3, -3)$ is on f^{-1}

$$\left[f^{-1}(x) \right]' = \frac{1}{f'(y)} \quad \text{so} \quad \left[f^{-1}(3) \right]' = \frac{1}{f'(-3)} = \frac{1}{-2} = -\frac{1}{2}$$

b) $m = \dfrac{-1}{2}$, pt. on inverse: $(3, -3)$

Tangent line: $y + 3 = \dfrac{-1}{2}(x - 3)$

TEST TIP

Study the steps of Example 34 carefully. You are almost guaranteed that a problem like this will appear on the AP test and it can be done in seconds if you understand the concept. This type of problem is a perfect candidate for the multiple-choice section as there are so many easy possibilities for answers for students who don't really comprehend inverses.

EXAMPLE 35: (Calculator Active) Find the derivative of $f^{-1}(x)$ for $f(x) = \ln x + e^{-x}$ at $x = 1$.

SOLUTION: $f(x) = \ln x + e^{-x}$

Inverse: $x = \ln y + e^{-y} = 1 \Rightarrow y \approx 2.505$

$$\frac{dy}{dx}_{(y=2.505)} = \frac{1}{\dfrac{1}{y} - e^{-y}} = \frac{1}{\dfrac{1}{2.505} - e^{-2.505}} \approx 3.149$$

Note that the answer can be found by plugging in your value of $y = 2.505$ in the derivative of the function and taking the reciprocal or you can use the NDERIV function on the calculator. Within three decimal places, the values are the same.

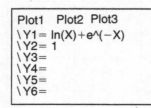

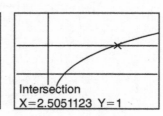

Horizontal and Vertical Tangent Lines

Overview: Horizontal and vertical tangent lines are important in function analysis and are covered in Chapter 6. However, since both are found using the derivative function, they are included here. In the figure below, two horizontal tangents and one vertical tangent to $f(x)$ are shown. In both cases, you need to work with $f'(x)$, the derivative of the function f. Express $f'(x)$ as a fraction.

Horizontal tangent lines: set $f'(x) = 0$, and solve for values of x in the domain of f (where the numerator of $f'(x) = 0$).

Vertical tangent lines: find values of x where $f'(x)$ is undefined in the domain of f (where the denominator of $f'(x) = 0$).

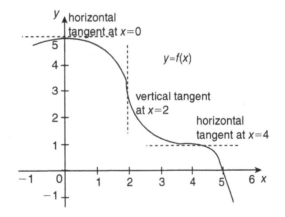

In both cases, to find the point of tangency, plug in the x-values you found into the function f. However, if both the numerator and the denominator of $f'(x)$ are simultaneously zero, no conclusion can be made about tangent lines. These types of problems go well with implicit differentiation.

EXAMPLE 36: Let $x^2 + x + y^4 - 4y = 17$. Find the equation of the line(s) that are vertically tangent to $f(x)$.

SOLUTION: $2x + 1 + (4y^3 - 4)\dfrac{dy}{dx} = 0 \Rightarrow \dfrac{dy}{dx} = \dfrac{-2x - 1}{4(y^3 - 1)}$

Vertical tangent when $y^3 - 1 = 0$ or $y = 1$

$x^2 + x + 1 - 4 = 17$

$x^2 + x - 20 = 0$

$(x + 5)(x - 4) = 0$

$x = -5, x = 4$

Note that this cannot be verified with a graphing calculator as solving for y is quite difficult. Below is the verification from a graphing utility that graphs implicitly.

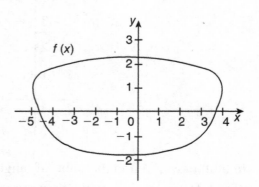

EXAMPLE 37: (Calculator Active) If $\sin x + \sin y = y$, $-\pi \le x \le \pi$, find all points of horizontal tangency to the curve.

SOLUTION:

$$\cos x + \cos y \frac{dy}{dx} = \frac{dy}{dx}$$

$$\frac{dy}{dx} = \frac{\cos x}{1 - \cos y}$$

Horizontal tangent when $\cos x = 0$ or $x = \frac{\pi}{2}, \frac{-\pi}{2}$

To find points, plug both values of x into the original equation:

$x = \frac{\pi}{2} : 1 + \sin y = y$ so $y = 1.935$ and point: $\left(\frac{\pi}{2}, 1.935 \right)$

$x = -\frac{\pi}{2} : -1 + \sin y = y$ so $y = -1.935$ and point: $\left(-\frac{\pi}{2}, -1.935 \right)$

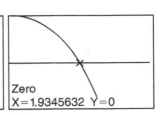

```
Plot1  Plot2  Plot3
\Y1 = 1+sin(X)−X
\Y2 =
\Y3 =
\Y4 =
\Y5 =
\Y6 =
\Y7 =
```

Zero
X=1.9345632 Y=0

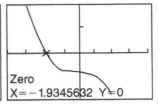

```
Plot1  Plot2  Plot3
\Y1 = −1+sin(X)−X
\Y2 =
\Y3 =
\Y4 =
\Y5 =

\Y7 =
```

Zero
X=−1.9345632 Y=0

TEST TIP

Read the problem! In Example 37, you were asked for points. You have x-values and you need y-values. You cannot solve equations like 1+ sin y = y analytically. But it can be solved graphically if there is only one variable in the equation. Put all variables to one side and set equal to zero. Put it into the calculator in terms of x. Graph it and find the x-intercept. Remember though that in this case, you are finding y, not x. This is a technique that you may have to know for the AP exam.

Differentiability

Overview: We have used the term *differentiable*. A **differentiable function** is one for which the derivative exists at all values in its domain. Functions that are not continuous at x-values cannot be differentiable at those x-values either. Typical AP questions that ask about differentiability are usually expressed giving piecewise functions.

Think of differentiability as "smooth." At the value c, where the piecewise function changes, the transition from one curve to another must be a smooth one. Sharp corners (like an absolute value curve) or cusp points (where two curves meet at a sharp point) mean the function is not differentiable there. Curves that have a vertical tangent at a point are also not differentiable at that point. The test for differentiability at $x = c$ is to show that $\lim_{x \to c^-} f'(x) = \lim_{x \to c^+} f'(x) = f'(c)$. So if you are given a piecewise function, check first for continuity at $x = c$, and if it is continuous, take the derivative of each piece, and check that the derivative is continuous at $x = c$. Lines, polynomials, exponentials, logarithms (within its domain), and sine and cosine curves are differentiable everywhere.

 EXAMPLE 38: Show that $f(x) = |x|$ is not a differentiable function.

SOLUTION: $f(x) = |x|$ is a V-shaped graph made up of two lines as shown in the figure. Each line is continuous and differentiable so the question is: Is the function differentiable at $x = 0$? It is clearly continuous there.

Writing $f(x) = |x|$ as a piecewise function, we get

$$f(x) = \begin{cases} x, x \ge 0 \\ -x, x < 0 \end{cases} \text{ so } f'(x) = \begin{cases} 1, x \ge 0 \\ -1, x < 0 \end{cases}$$

$\lim_{x \to 0^-} f'(x) = -1$ and $\lim_{x \to 0^+} f'(x) = 1$ so $\lim_{x \to 0} f'(x)$ does not exist and $f(x)$ is not differentiable at $x = 0$.

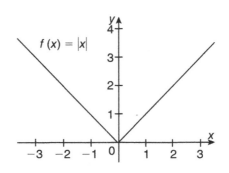

EXAMPLE 39: Let $f(x) = \begin{cases} -2x^2 + 6x - 1, x \le 2 \\ \cos(4x - 8) - 2x + 6, x > 2 \end{cases}$. Show that $f(x)$ is differentiable.

SOLUTION: Both a parabola and a cosine curve are continuous. $f(x)$ is continuous at $x = 2$ because $\lim_{x \to 2^-} f(x) = -8 + 12 - 1 = 3$

$\lim_{x \to 2^+} f(x) = \cos 0 - 4 + 6 = 3$

so $\lim_{x \to 2} f(x) = 3$

$f'(x) = \begin{cases} -4x + 6, x \le 2 \\ -4\sin(4x - 8) - 2, x > 2 \end{cases}$

$f(x)$ is differentiable at $x = 2$ because

$\lim_{x \to 2^-} f'(x) = -8 + 6 = -2$

$\lim_{x \to 2^+} f'(x) = -4\sin 0 - 2 = -2$

so $\lim_{x \to 2} f'(x) = -2$

The graph of the function is below. Notice the smooth transition at $x = 2$.

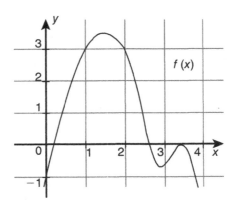

EXAMPLE 40: If $f(x) = \begin{cases} \dfrac{ax^3}{3} - x^2 + 3, & x \geq -1 \\ \dfrac{b(x+3)^2}{2} - 2x, & x < -1 \end{cases}$, find the values of a and b that make f differentiable.

SOLUTION: $\displaystyle\lim_{x \to -1^-} f(x) = 2b + 2 \qquad \lim_{x \to -1^+} f(x) = \dfrac{-a}{3} - 1 + 3 = \dfrac{-a}{3} + 2$

Continuity: $2b + 2 = \dfrac{-a}{3} + 2$ so $a = -6b$

$f'(x) = \begin{cases} ax^2 - 2x, & x \geq -1 \\ b(x+3) - 2, & x < -1 \end{cases}$

$\displaystyle\lim_{x \to -1^-} f'(x) = 2b - 2 \qquad \lim_{x \to -1^+} f'(x) = a + 2$

Differentiability: $2b - 2 = a + 2$ so $a = 2b - 4$

Since $a = -6b$, $2b - 4 = -6b$ so $b = \dfrac{1}{2}$ and $a = -6b = -3$

$a = -3, b = \dfrac{1}{2}$

The graphical verification is below.

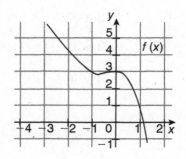

EXAMPLE 41: The graph of $f(x) = 1.01 - \sqrt{(x+1)^2 + .0001}$ is shown in the figure below. If this function is differentiable, explain why. If it is not, explain why not.

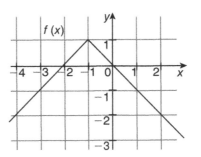

SOLUTION: The function is continuous at $x = -1$. But is it differentiable there?

The graph has what appears to be a corner at $x = -1$, but looks are deceiving.

$f'(x) = \dfrac{-(x+1)}{\sqrt{(x+1)^2 + .0001}}$ and $f'(-1) = 0$. Thus the function is

differentiable.

TEST TIP

Always view given graphs on the AP exam with skepticism. While a graph may give you an idea of the general shape and behavior of the function, it cannot always show the detail necessary to make conclusions. The function given algebraically is always preferable.

Time for a quiz
- Review strategies in Chapter 2
- Take Quiz 3 at the REA Study Center
  (www.rea.com/studycenter)

Applications of Differentiation

Related Rates

Overview: Related-rates problems are sometimes difficult for students as they contain a lot of words and, typically, a lot of information is crammed into several sentences. All of this information can be confusing, so it is important to understand the kind of information you will be given.

If Ted's height is given to be 5 feet 6 inches, for instance, you would write that as we have done throughout algebra: choose a variable for height (probably h), convert to inches, and simply state $h = 66$. However, if you are told that Ted is growing at the rate of 2 inches/year, you must realize that this is a rate and rates are written as derivatives with respect to time. This information would be written as $\frac{dh}{dt} = 2$. If you were told that Jen is on a diet and losing 3 pounds a week, you would probably use the variable w for weight and write $\frac{dw}{dt} = -3$.

Related rates questions can always be identified by words like: *increasing, decreasing, growing, shrinking, expanding, descending*, and so on.

TEST TIP

It is possible that this rate of growth or decline can change as well. Since rate of growth is a derivative and change also means a derivative, the rate of growth changing refers to a second derivative. For instance, if you were told that a tree was 20 feet tall, growing at the rate of 4 feet/year, and this rate of growth is slowing by 0.5 feet/year/year, you would write this information as: $h = 20, \quad \dfrac{dh}{dt} = 4, \quad \dfrac{d^2h}{dt^2} = -0.5$.

EXAMPLE 1: Translate the following statements into algebraic equations:

(a) The temperature of an oven is 350°F and increasing at the rate of $\dfrac{15°F}{\min}$.

(b) The amount of ink in the reservoir of my printer is 0.5 in³ and reducing at the rate of $0.1 \dfrac{\text{in}^3}{\text{hr}}$.

(c) When I have done 30 sit-ups, I can do 20 sit-ups/minute and this rate is decreasing by 5 sit-ups per minute per minute.

SOLUTIONS:

(a) $T = 350, \dfrac{dT}{dt} = 15$

(b) $V = 0.5, \dfrac{dV}{dt} = -0.1$ (cubic units indicates information about volume)

(c) $s = 30, \dfrac{ds}{dt} = 20, \dfrac{d^2s}{dt^2} = -5$

DIDYOU**KNOW?**

"In the fall of 1972, President Nixon announced that the rate of increase of inflation was decreasing. This was the first time a sitting president used the third derivative to advance his case for reelection." – Hugo Rossi, Professor Emeritus of Math, University of Utah

EXAMPLE 2: If the area of an unmowed lawn is a function of time given by $A(t)$, A in ft^2 and t in minutes, explain in words the meaning of $\dfrac{dA}{dt} = -12$. Specify units.

SOLUTION: The area of the unmowed lawn is decreasing by 12 ft^2 per minute (or the lawn is being mowed at the rate of 12 ft^2 per minute).

Procedure for Solving Related-Rates Problems

(1) If no diagram is given, make one if it helps to clarify the problem. Label all parts in terms of variables even if you are given the actual values of the variable.

(2) Make a list of all variables that are given and those that are asked for. Be aware of variables that are constants (never changing) and variables that change over time. Be sure that the variables you use in your list match those used in the diagram. Remember that rates are derivatives with respect to time t. Pay careful attention to words that indicate whether these rates are positive or negative.

(3) Find an equation that ties your variables together. Right triangle problems typically use the Pythagorean Theorem. Area problems typically use geometric formulas that you know. Problems with angles typically use trigonometric functions. Usually, any formula that models volume and surface area will be given to you.

(4) You may (and should) plug into this formula the value of any variable that is a constant and not changing. **Never** plug in any variable that is not a constant at this time.

(5) Differentiate your equation with respect to time. This is implicit differentiation with respect to t.

(6) You are now free to plug in all variables. If you have more than one unknown variable, you need an equation to find the value of an unknown variable at that specific time. Usually, but not always, it is the same equation as the one you use in step 3.

(7) Solve your original equation by solving for the variable for which the problem asks. Be sure to label it with correct units.

EXAMPLE 3: A spherical balloon is being blown up at the rate of $3\dfrac{in^3}{sec}$. When the diameter of the balloon is 4 inches, a) determine how fast the diameter is changing and b) determine how fast the surface area of the balloon is changing. Specify units. (The volume of a sphere is $V = \dfrac{4}{3}\pi r^3$ and the surface area is $S = 4\pi r^2$.)

SOLUTIONS: No diagram is necessary.

Given: $\dfrac{dV}{dt} = 3$, $r = 2$, and we want $\dfrac{dd}{dt}$ and $\dfrac{dS}{dt}$.

(a) $\dfrac{dV}{dt} = 4\pi r^2 \dfrac{dr}{dt}$

$3 = 4\pi(2^2)\dfrac{dr}{dt}$

$\dfrac{dr}{dt} = \dfrac{3}{16\pi}$

The diameter is changing at the rate of $\dfrac{3\ in}{8\pi\ min}$.

(b) $\dfrac{dS}{dt} = 8\pi r \dfrac{dr}{dt}$

We use the $\dfrac{dr}{dt}$ found in part a). Note that even if part a) was not asked, it still needs to be done.

$\dfrac{dS}{dt} = 8\pi(2)\dfrac{3}{16\pi} = 3$

The surface area is changing at the rate of $3\dfrac{in^2}{min}$.

EXAMPLE 4: Ike and Mike are biking along country roads that are perpendicular to each other. At the intersection of the roads is a stoplight. Ike is 2 miles north of the intersection and traveling north at 12 mph. Mike is 3 miles east of the intersection traveling at 10 mph. How fast is the distance between them changing if Mike is traveling a) east and b) west?

SOLUTION: Given: a (Ike) $= 2$, $\dfrac{da}{dt} = 12$, b (Mike) $= 3$, $\dfrac{db}{dt} = \pm10$. Find $\dfrac{dc}{dt}$.

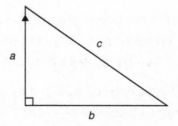

$$c^2 = a^2 + b^2$$

$$2c\frac{dc}{dt} = 2a\frac{da}{dt} + 2b\frac{db}{dt}$$

For part a), $\frac{db}{dt} = 10$ because Mike's distance from the intersection is increasing.

$2c\frac{dc}{dt} = 2(2)(12) + 2(3)(10)$. There are two unknowns but at this moment in time, $c = \sqrt{13}$ by the Pythagorean Theorem.

$$2\sqrt{13}\frac{dc}{dt} = 108$$

$$\frac{dc}{dt} = \frac{54}{\sqrt{13}}$$

The bikers are separating at the rate of $\frac{54}{\sqrt{13}}$ mph.

For part b), $\frac{db}{dt} = -10$ because Mike's distance from the intersection is decreasing.

$$2\sqrt{13}\frac{dc}{dt} = 2(2)(12) + 2(3)(-10)$$

$$\frac{dc}{dt} = \frac{-6}{\sqrt{13}}$$

The bikers are getting closer at the rate of $\frac{6}{\sqrt{13}}$ mph.

TEST TIP

Pay attention to words that signify change. Signs are important and can mean the difference between receiving credit or not on an AP exam question. And even if you have no idea how to solve a related-rates problem, just writing changing units as derivatives with correct signs and units can get you a great deal of partial credit on the exam.

EXAMPLE 5: A paper cup in the shape of a cone has a top diameter of 8 inches and a height of 8 inches. The cup has water in it and the water is draining from the bottom at the rate of $\dfrac{2 \text{ in}^3}{\text{min}}$. Determine how fast the height of the water is changing when the water in the cup is 6 inches high. The volume of a cone is given by $V = \dfrac{1}{3}\pi r^2 h$.

SOLUTION: First it is important to realize that the values of the radius r and the height h are not given. These are the dimensions of the cup. We must concentrate on what is actually changing, which is the height of the water. We are given: $h = 6$ and $\dfrac{dV}{dt} = -2$ and we want $\dfrac{dh}{dt}$. Taking the derivative of V would involve the product rule as there are two variables on the right side of the equation. We need a relationship between r and h. The second diagram illustrates that we can set up a relationship with similar triangles: $\dfrac{r}{h} = \dfrac{4}{8}$. We can solve for either variable but it makes sense to solve for r in terms of h as we have information about h and we want $\dfrac{dh}{dt}$. So here is the solution:

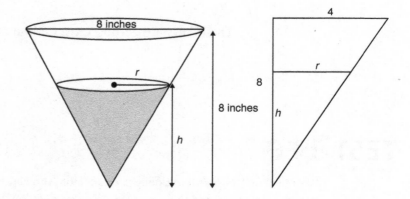

$$8r = 4h \text{ so } r = \frac{h}{2}$$

$$V = \frac{1}{3}\pi r^2 h = \frac{1}{3}\pi\left(\frac{h}{2}\right)^2 h = \frac{\pi h^3}{12}$$

$$\frac{dV}{dt} = \left(\frac{dV}{dh}\right)\left(\frac{dh}{dt}\right)$$

$$\frac{dV}{dt} = \frac{\pi h^2}{4}\frac{dh}{dt}$$

$$-2 = \frac{36\pi}{4}\frac{dh}{dt}$$

$$\frac{dh}{dt} = \frac{-2}{9\pi}$$

The height is decreasing at the rate of $\dfrac{2\,\text{in}}{9\pi\,\text{min}}$.

EXAMPLE 6: A lighthouse is 4 miles from point P along a straight shoreline and the light from this lighthouse makes 4 revolutions per minute. How fast, in miles per minute is the beam of light moving along the shoreline when it is 3 miles from point P?

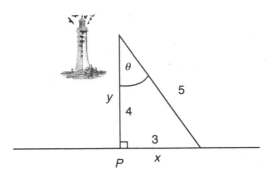

SOLUTION: In the diagram, label the important variables as x, y, and θ. We are given that $y = 4$ (constant), $x = 3$, and $\dfrac{d\theta}{dt} = \dfrac{4\,\text{rev}}{\text{min}}$. You want $\dfrac{dx}{dt}$. Since we are given information about angles, you want a trigonometric function using x and y. Our equation would be $\tan\theta = \dfrac{x}{y}$. We can immediately plug in the value of y because it is a constant so we get $\tan\theta = \dfrac{x}{4}$. Taking the derivative, we get $\sec^2\theta\,\dfrac{d\theta}{dt} = \dfrac{1}{4}\dfrac{dx}{dt}$. It is important to understand that $\dfrac{d\theta}{dt}$ must be measured using radians so we get

$\dfrac{d\theta}{dt} = \dfrac{4 \text{ rev}}{\text{min}}\left(\dfrac{2\pi}{\text{rev}}\right) = \dfrac{8\pi}{\text{min}}$. And although we do not know the value of

θ, we know that the hypotenuse of the triangle is 5 so $\sec\theta = \dfrac{5}{4}$. Putting it together:

$$\sec^2\theta \dfrac{d\theta}{dt} = \dfrac{1}{4}\dfrac{dx}{dt}$$

$$\left(\dfrac{5}{4}\right)^2 (8\pi) = \dfrac{1}{4}\dfrac{dx}{dt}$$

$$\dfrac{dx}{dt} = 50\pi \ \dfrac{\text{miles}}{\text{min}}$$

TEST TIP

Whenever angles are used in a calculus formula, they must be expressed in radians. Whenever angles are found by using a formula, they are automatically calculated in radians. If rates are given in degrees/time or revolutions/time, they must be converted. To change angles to radians, use this relationship: $2\pi = 360° = 1$ revolution. In multiple-choice problems on the AP exam, it is almost guaranteed that one or more of the incorrect answers will involve all of the correct calculations, but use degrees instead of radians.

Straight-Line Motion

Overview: Straight-line motion problems typically refer to a particle traveling along the x-axis or y-axis. Its position on that axis is a function of time and usually is referred to as $x(t)$ or $y(t)$. In these problems, time t is just about always greater than or equal to 0.

DID YOU KNOW?

Calculus has been used in astronomy since the 17th century to calculate the orbits of the planets around stars. Calculus is also necessary to accurately calculate the changing speeds of moving objects in space, including asteroids, comets, planets, and solar systems. Our knowledge of the universe today would be insignificant without calculus.

The **instantaneous rate of change** of $x(t)$ or $y(t)$ is a derivative and refers to the **velocity** of the particle. This is usually referred to as $v(t)$. So in the case of a particle moving along the x-axis, $v(t) = \dfrac{dx}{dt} = x'(t)$ and in the case of a particle moving along the y-axis, $v(t) = \dfrac{dy}{dt} = y'(t)$. The velocity function not only tells you how fast the particle is moving at any time t; it tells you its direction according to the following chart.

	horizontal movement	vertical movement
$v(t) > 0$	right	up
$v(t) < 0$	left	down
$v(t) = 0$	stopped	stopped

The velocity function is measured in $\dfrac{\text{linear units}}{\text{time}}$. Examples are $\dfrac{-20 \text{ meters}}{\text{sec}}$ and 45 mph.

Speed is not synonymous with velocity. Speed is always positive. We therefore write that speed $= |v(t)|$. When a particle is always moving right horizontally or always moving upwards vertically, we can say that speed $= v(t)$.

Acceleration is the instantaneous rate of change of velocity and thus can be expressed as a derivative: $a(t) = v'(t)$. So, in the case of horizontal motion, $(a(t) = v'(t) = x''(t))$ and in the case of vertical motion, $a(t) = v'(t) = y''(t)$. The acceleration tells you how fast the particle's velocity is changing at any time t as well as the direction of that change. For example, when you hit the brakes while traveling forward, you apply a negative acceleration. Your velocity is still positive, but smaller than it was before.

	horizontal movement	vertical movement
$a(t) > 0$	accelerating right	accelerating up
$a(t) < 0$	accelerating left	accelerating down
$a(t) = 0$	no acceleration	no acceleration

No acceleration ($a(t) = 0$) does not mean that the particle is stopped. It means that its velocity is not changing. Cruise control on a car set to 50 mph is an example of $a(t) = 0$. The velocity would be a constant 50 mph with no acceleration or deceleration.

Whether a particle is speeding up or slowing down is based on *both* the sign of its velocity and acceleration at any time t. The following chart describes this.

	$a(t) > 0$	$a(t) < 0$
$v(t) > 0$	particle speeding up	particle slowing down
$v(t) < 0$	particle slowing down	particle speeding up

We can summarize this chart by saying: when the velocity and acceleration have the same sign at any time t, the particle is speeding up and when the velocity and acceleration have opposite signs at any time t, the particle is slowing down.

TEST TIP

Beware of interpreting a problem that states that the acceleration is always positive as a particle is always speeding up. This is only true when the particle's velocity is always positive as well. Information about the acceleration alone gives you insufficient information as to when the particle is speeding up.

EXAMPLE 7: A locomotive in a train yard that is traveling along a straight track has velocity $v(t)$, t measured in minutes, as described in the following figure. How many times does the locomotive change direction over 10 minutes?

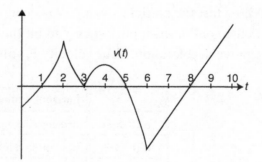

SOLUTION: Direction changes when $v(t)$ switches from positive to negative or negative to positive. This occurs at $t = 1$, $t = 5$, and $t = 8$. Note that at $t = 3$, the locomotive stops, but does not change direction. The locomotive changes direction three times.

EXAMPLE 8: A particle is moving along the *x*-axis and has velocity function $v(t)$ for $0 \le t \le 6$ as described in the following figure. For what values of *t* is the particle speeding up? Justify your answer.

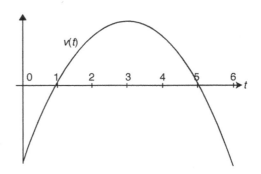

SOLUTION: $v(t) > 0$ for $1 < t < 5$ and $v(t) < 0$ for $0 \le t < 1$ and $5 < t \le 6$.

The acceleration function $a(t)$ is the derivative of $v(t)$ and thus the slope of $v(t)$. So,

$a(t) > 0$ for $0 \le t < 3$.
$a(t) < 0$ for $3 < t \le 6$.

The situation above can be described with sign charts.

$$v(t): \underline{-----0+++++++++++++0------}$$
$$\;0 \qquad\;\; 1 \qquad\qquad\qquad\quad 5 \qquad\quad 6$$

$$a(t): \underline{+++++++++++++0---------------}$$
$$\;0 \qquad\qquad\qquad\quad 3 \qquad\qquad\qquad 6$$

$v(t)$ and $a(t)$ have the same signs for $1 < t < 3$ and $5 < t \le 6$.

Thus the particle is speeding up on those intervals.

TEST TIP

Sign charts by themselves are never a justification on the AP exam. While you are encouraged to create them for your own benefit, on the exam you must describe in words what the sign chart says. In the above case, the explanation below the sign chart is the justification you would need to write.

The AP exam graders rarely get hung up on distinctions between the use of a greater than sign (>) or greater than or equal sign (≥). In the example above, technically both the velocity and acceleration are negative at $t = 6$ necessitating $5 < t \le 6$. However, full credit would be given for $5 < t < 6$.

EXAMPLE 9: A particle is moving along the y-axis such that any time $t \geq 0$, its position is given by $y(t) = 5 - 72t + 33t^2 - 4t^3$. For what values of t is the particle moving down?

SOLUTION: $v(t) = y'(t) = -72 + 66t - 12t^2 = -6\left(2t^2 - 11t + 12\right)$

$v(t) = -6(2t - 3)(t - 4) = 0$

$t = \dfrac{3}{2}, t = 4$

$v(t): \underline{-----0+++++++++++++++0------}$
$\qquad\quad 0 \qquad\;\; 3/2 \qquad\qquad\qquad\quad 4$

$v(t) < 0$ on $\left[0, \dfrac{3}{2}\right)$ and $(4, \infty)$

EXAMPLE 10: A particle is moving along the x-axis in such a way that at any time $t \geq 0$ its position is given by $x(t) = \dfrac{\ln t}{t}$. What is the particle's acceleration at $t = e$?

SOLUTION:

$v(t) = \dfrac{t\left(\dfrac{1}{t}\right) - \ln t}{t^2} = \dfrac{1 - \ln t}{t^2}$

$a(t) = \dfrac{t^2\left(-\dfrac{1}{t}\right) - (1 - \ln t)(2t)}{t^4} = \dfrac{-t - 2t(1 - \ln t)}{t^4} = \dfrac{-t(1 + 2 - 2\ln t)}{t^4} = \dfrac{2\ln t - 3}{t^3}$

$a(e) = \dfrac{2 - 3}{e^3} = \dfrac{-1}{e^3}$

EXAMPLE 11:   A billboard has a light that goes up and down in a sporadic pattern. The velocity of the light is given by the following table in 2-second intervals: Find an approximation using three different calculations for the acceleration of the light at $t = 10$ seconds. Express units.

t (seconds)	$v(t)$ ft/second
0	4
2	6
4	2
6	0
8	−1
10	−4
12	−2
14	0
16	−1
18	1
20	3

SOLUTION: $a(10) \approx \dfrac{v(10) - v(8)}{10 - 8} = \dfrac{-4 + 1}{2} = \dfrac{-3}{2} \dfrac{\text{ft}}{\text{sec}^2}$ OR

$a(10) \approx \dfrac{v(12) - v(10)}{12 - 10} = \dfrac{-2 + 4}{2} = 1 \dfrac{\text{ft}}{\text{sec}^2}$ OR

$a(10) \approx \dfrac{v(12) - v(8)}{12 - 8} = \dfrac{-2 + 1}{4} = \dfrac{-1}{4} \dfrac{\text{ft}}{\text{sec}^2}$

TEST TIP

When you are asked on the AP exam to find an approximation, be sure to use the approximation sign ($\approx$) rather than an equal sign. You will not receive full credit if you use an equal sign.

EXAMPLE 12: Newton the cat is walking along a windowsill such that at any time $t \geq 0$ his velocity is given by $v(t) = \dfrac{\sin t}{\sin t - \cos t + 2}$. Determine whether Newton is speeding up or slowing down at $t = \dfrac{\pi}{2}$.

SOLUTION:

$$v\left(\frac{\pi}{2}\right) = \frac{1}{1-0+2} = \frac{1}{3}$$

$$a(t) = \frac{(\sin t - \cos t + 2)\cos t - \sin t(\cos t + \sin t)}{(\sin t - \cos t + 2)^2}$$

$$a\left(\frac{\pi}{2}\right) = \frac{(1-0+2)(0) - 1(0+1)}{(1-0+2)^2} = \frac{-1}{9}$$

Since v and a have opposite signs at $t = \dfrac{\pi}{2}$, Newton is slowing down.

TEST TIP

Students frequently confuse the values of quadrant angles for the sine and cosine functions. Many non-calculator questions on the AP exam use quadrant angles. You could memorize the values of the trig functions at quadrant angles using this chart, but the best way to remember them is by their graphs shown below:

θ	$\sin \theta$	$\cos \theta$
0	0	1
$\pi/2$	1	0
π	0	-1
$3\pi/2$	-1	0
2π	0	1

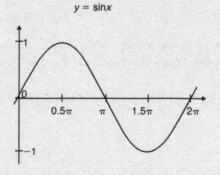
$y = \sin x$

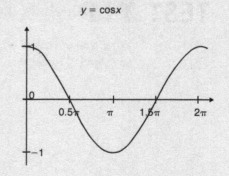
$y = \cos x$

EXAMPLE 13: (Calculator Active) A particle moves along a straight line such that its velocity at any time $t \geq 0$ is given by $v(t) = \dfrac{\cos t - e^{2t}}{\ln(t+1) + \sqrt{t^2 + 2^t}}$. Determine whether the particle is speeding up or slowing down at $t = 1$.

SOLUTION: This problem has everything in it but the kitchen sink and, while finding the acceleration function algebraically by taking its derivative is possible, it is time-consuming. And besides, you are still going to have to plug the value of 1 into the given function and the derivative. So use the calculator to find the value of the given function at $t = 1$ and its derivative at $t = 1$.

```
Plot1 Plot2 Plot3
\Y₁ =(cos(X)−e^(2
X))/((ln(X+1)+√(
X²+2^X))
\Y₂ =
\Y₃=
\Y₄=
\Y₅=
```

```
Y₁(1)
            −2.824
nDeriv(Y₁, X, 1)
            −4.720
```

Since $v(1) < 0$ and $a(1) < 0$, the particle is speeding up at $t = 1$.

TEST TIP

The danger in using the calculator is that functions need to be input correctly. A missing or misplaced parenthesis can make a difference and create an altogether different function. Students typically believe whatever the calculator says, dismissing the possibility of a function input incorrectly. The AP exam developers know the types of mistakes that students make when inputting equations into the calculator and include these answers as incorrect choices. So be careful! If it is not too time-consuming, problems should be done algebraically and the calculator used for verification. In the case of the problem above, though, you are being tested on your ability to use the calculator and taking the derivative algebraically would be a huge waste of time.

The Intermediate Value Theorem

Overview: The **Intermediate Value Theorem (IVT)** states that when we have a continuous function f on the closed interval $[a, b]$ where $f(a) \neq f(b)$, the function must take on every value between $f(a)$ and $f(b)$ at some point between a and b. In the diagram, we have a continuous curve f connecting one point $(a, f(a))$ below the line and the other point $(b, f(b))$ above the line. The IVT states that there must be at least one point where the curve crosses the line. For instance, if at 8 a.m. the outside temperature was 60°F and at 4 p.m. the outside temperature was 80°F, there must have been at least one time between 8 a.m. and 4 p.m. that the outside temperature was 75°F. Certainly that could have occurred numerous times, but it must have happened at least once. The IVT is essentially a pre-calculus concept, but it has shown up on AP exams and students should know it by name.

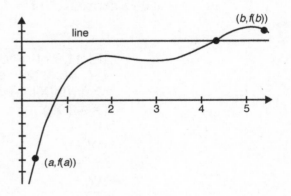

EXAMPLE 14: A conveyor belt has the ability to go both forwards and backwards. During the time interval $0 \leq t \leq 60$ seconds, the velocity function of the belt $v(t)$, measured in inches/sec is a continuous function. The table below shows values of this function at selected times. For $0 \leq t \leq 60$, what is the minimum number of times that the belt must have been stopped? Justify your answer.

t (sec)	0	10	20	30	40	50	60
v (t) (in/sec)	6	8	−3	−3	0	−5	−2

SOLUTION: There are at least two times that the belt must have been stopped … $v(t) = 0$. The Intermediate Value Theorem applies because the velocity function is continuous. Since $v(10) = 8 > 0 > -3 = v(20)$, the IVT guarantees a time t on $(10, 20)$ such that $v(t) = 0$. And, of course, $v(t) = 0$ when $t = 40$.

EXAMPLE 15: The functions f and g are continuous. The continuous function h is given by $h(x) = -x \cdot g\,(f\,(x))$. The table below gives selected values of the functions. Explain why there must be a value t for $1 < t < 7$ such that $h(t) = -10$.

x	1	3	5	7
f (x)	5	2	-3	1
g (x)	2	-4	-1	3

SOLUTION: $h(1) = -1 \cdot g\,(f\,(1)) = -1 \cdot g\,(5) = -1(-1) = 1$
$h(7) = -7 \cdot g\,(f\,(7)) = -7 \cdot g\,(1) = -7(2) = -14$
Since $h(7) < -10 < h(1)$, and h is continuous, by the IVT there exists a value t on $(1, 7)$ such that $h(t) = -10$.

The Mean-Value Theorem and Rolle's Theorem

Overview: The **Mean-Value Theorem (MVT)** is one of the few theorems in calculus that you must know by name. To complicate matters, there is a second MVT in the integration section of the course. However, the one in this section is much more important.

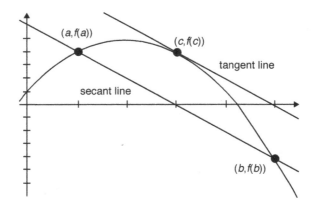

The formal definition of the MVT says that if a function f is continuous on some interval $[a, b]$ and differentiable on the interval (a, b), there exists some c, $a < c < b$, such that $f'(c) = \dfrac{f(b) - f(a)}{b - a}$. What the MVT states is that there must be some value of c between a and b such that the tangent line to f at $x = c$ is parallel to the secant line containing $(a, f(a))$ and $(b, f(b))$. This is illustrated in the figure above.

An example of the Mean-Value Theorem is this: if the average temperature during a day is 60 degrees, at some point during the day the temperature *had to be* 60 degrees. Or, while driving on a road, if you average 65 mph between 1 p.m. and 2 p.m., there must be at least one time between 1 p.m. and 2 p.m. when you were actually traveling at 65 mph.

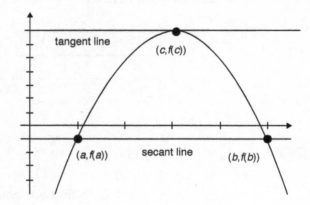

A special case of the mean-value theorem is **Rolle's Theorem**. This adds the stipulation that $f(a) = f(b)$. In this case, the secant line containing the two given points is horizontal; therefore the tangent line at some point c between a and b must also be horizontal and its slope equals zero: $f'(c) = 0$.

EXAMPLE 16: Show that Rolle's Theorem holds between the roots of $f(x) = 12x - x^3$.

SOLUTION: $f(x)$ is a polynomial and thus continuous and differentiable everywhere, as shown in the figure below.

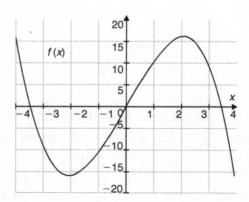

$$12x - x^3 = 0$$
$$x(12 - x^2) = 0$$
Roots: $x = 0$, $x = \pm 2\sqrt{3}$

$$f'(x) = 12 - 3x^2 = 0$$
$$3x^2 = 12$$
$$x = \pm 2$$

At $x = 2$ and $x = -2$ the tangent lines to f are horizontal.

EXAMPLE 17: (Calculator Active) Find all values of x that satisfy the Mean-Value Theorem for $f(x) = \sin x + 2\cos 2x$ on $\left[0, \dfrac{\pi}{2}\right]$.

SOLUTION: The MVT holds because sine and cosine curves are continuous and differentiable.

$$\text{MVT: } f'(x) = \frac{f\left(\dfrac{\pi}{2}\right) - f(0)}{\dfrac{\pi}{2} - 0} = \frac{(1-2) - (0+2)}{\dfrac{\pi}{2}}$$

$$\cos x - 4\sin 2x = \frac{-6}{\pi}$$

$$x = 0.394, \ x = 1.28$$

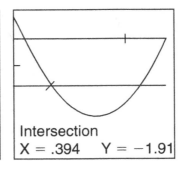

```
Plot1  Plot2  Plot3
\Y₁ =cos(X)−4sin(
2X)
\Y₂ = −6/π
\Y₃=
\Y₄=
\Y₅=
\Y₆=
```

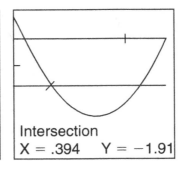

Intersection
X = .394   Y = −1.91

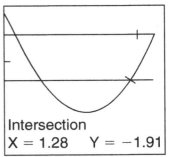

Intersection
X = 1.28   Y = −1.91

TEST TIP

The last two problems were straightforward. You were told exactly what to do. While these types of problems occur on the AP exam, far more frequent are the problems that ask you to explain why some calculus fact is so and you have to utilize theorems like the MVT without being specifically told to do so. Examples 18 and 19 illustrate this.

EXAMPLE 18: A ball is bobbing along in the ocean and traveling a straight line due to the waves. Its velocity over a 60 second period is a differentiable function given at 6-second intervals in the following table.

t (sec)	0	6	12	18	24	30	36	42	48	54	60
v (t) ft/sec	5	2	−2	0	−3	−5	−1	3	4	−2	−4

a) What is the minimum number of times the ball is stopped?
b) Between what two times must the ball have no acceleration? Explain.

SOLUTIONS: (a) Because of the Intermediate Value Theorem, the ball must be stopped ($v(t) = 0$) on [6, 12], at $t = 18$, at some time on [36, 42] and [48, 54]. The ball must be stopped a minimum of four times.

(b) The ball must have no acceleration ($a(t) = 0$) at some time on [12, 54] because of the Mean Value Theorem:

$$v'(t) = a(t) = \frac{v(54) - v(12)}{54 - 12} = \frac{-2 + 2}{42} = 0$$

Note that this is actually an example of Rolle's Theorem.

EXAMPLE 19: Let f be a twice-differentiable function. Selected values of f and f' are given in the table below. Let h be the function given by $h(x) = f(f(x))$. Show that there must be a value c, $1 < c < 4$, such that $h''(c) = -10$.

x	1	2	3	4
$f(x)$	3	0	−4	2
$f'(x)$	−1	−4	−2	7

SOLUTION: The MVT states: $h''(c) = \dfrac{h'(4) - h'(1)}{4 - 1}$

Because $h'(x) = f'(f(x)) f'(x)$

$$h''(c) = \frac{f'(f(4))f'(4) - f'(f(1))f'(1)}{3}$$

$$h''(c) = \frac{(f'(2))(7) - (f'(3))(-1)}{3}$$

$$h''(c) = \frac{(-4)(7) - (-2)(-1)}{3} = \frac{-30}{3} = -10$$

Function Analysis

Overview: Prior to 1995, AP calculus exams typically asked students to graph functions. With the advent of graphing calculators, it is rare that students are asked to do so in today's exams. Rather, they are asked to analyze graphs of functions given its equation or some aspect of the function. Function analysis is probably the most important aspect of the AP calculus exam as so many problems depend on its techniques.

In pre-calculus, you were taught how to find **roots** of functions. Roots are the x-intercepts. While the AP exam requires you to know the techniques of finding roots by setting the function equal to zero, most problems involve calculus techniques to analyze functions in greater depth.

Given a function $y = f(x)$, the goal is to find intervals where $f(x)$ is increasing, decreasing, concave up, and concave down. Also, you can be asked to find values of x where the function has points of relative maximum, relative minimum, and inflection.

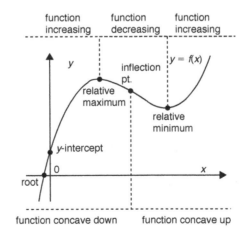

Critical values: values of $x = c$ where $f'(c) = 0$ or $f'(c)$ is undefined.

Increasing/Decreasing

$f(x)$ is **increasing** for values of k such that $f'(k) > 0$. And if $f'(k) > 0$, then f is increasing at $x = k$.

$f(x)$ is **decreasing** for values of k such that $f'(k) < 0$. And if $f'(k) < 0$, then f is decreasing at $x = k$.

Concavity

$f(x)$ is concave up for values of k such that $f''(k) > 0$. And if $f''(k) > 0$ for all k on (a, b), then f is concave up on (a, b).

$f(x)$ is concave down for values of k such that $f''(k) < 0$. And if $f''(k) < 0$ for all k on (a, b), then f is concave down on (a, b).

If $f(x)$ switches concavity at $x = c$, then it has a **point of inflection** at $x = c$. If $f''(c) = 0$, that does not necessarily indicate that f has a point of inflection at $x = c$.

Relative (or Local) Extrema

If $f(x)$ switches from increasing to decreasing at $x = c$, alternately $f'(x)$ switches from positive to negative at $x = c$, the function f has a **relative (or local) maximum** at $x = c$.

If $f(x)$ switches from decreasing to increasing at $x = c$, alternately $f'(x)$ switches from negative to positive at $x = c$, the function f has a **relative (or local) minimum** at $x = c$.

These steps are commonly known as the **first derivative test.**

Relative extrema can also be found using the **second derivative test**.

If $f'(k) = 0$ and $f''(k) > 0$, $f(x)$ has a relative minimum at $x = k$.

If $f'(k) = 0$ and $f''(k) < 0$, $f(x)$ has a relative minimum at $x = k$.

If $f'(k) = 0$ and $f''(k) = 0$, the test is inconclusive.

EXAMPLE 20: Given $f(x) = \dfrac{x^2 - 1}{x^3}$.

  a) Find the x-coordinate of all critical points of f and determine whether each is a relative minimum or relative maximum.
  b) Find intervals where the graph of f is concave up.
  c) Find points of inflection. Justify all answers.

SOLUTIONS: (a) $f(x) = \dfrac{x^2 - 1}{x^3} = \dfrac{x^2}{x^3} - \dfrac{1}{x^3} = x^{-1} - x^{-3}$

$f'(x) = -1x^{-2} + 3x^{-4} = \dfrac{-1}{x^2} + \dfrac{3}{x^4}$

$f'(x) = \dfrac{-1}{x^2}\left(\dfrac{x^2}{x^2}\right) + \dfrac{3}{x^4} = \dfrac{3 - x^2}{x^4} = 0$

$3 - x^2 = 0 \Rightarrow x = \pm\sqrt{3}$

Also, $x = 0$ is a critical value but the function is undefined there.

$f'(x): \underline{----- 0 +++\infty +++ 0 -----}$
$ -\sqrt{3} \quad\quad 0 \quad\quad \sqrt{3}$

f has a relative maximum at $x = \sqrt{3}$ because f' switches from positive to negative.

f has a relative minimum at $x = -\sqrt{3}$ because f' switches from negative to positive.

(b) $f''(x) = 2x^{-3} - 12x^{-5} = \dfrac{2}{x^3} - \dfrac{12}{x^5}$

$f''(x) = \dfrac{2}{x^3}\left(\dfrac{x^2}{x^2}\right) - \dfrac{12}{x^5} = \dfrac{2x^2 - 12}{x^5} = 0$

$2x^2 - 12 = 0 \Rightarrow x = \pm\sqrt{6}$

$f''(x): \underline{---- 0 +++\infty +++ 0 ----}$
$ -\sqrt{6} \quad\quad 0 \quad\quad \sqrt{6}$

f is concave up on $\left(-\sqrt{6}, 0\right)$ and $\left(0, \sqrt{6}\right)$ because $f''(x) > 0$. Note that the answer $\left(-\sqrt{6}, \sqrt{6}\right)$ is incorrect as $x = 0$ is not in the domain of f.

(c) There are inflection points at $x = \pm\sqrt{6}$ only, as $f''(x)$ changes sign at these values.

EXAMPLE 21: The figure below shows the graph of $y = f'(x)$, the derivative of the function f for $-4 \leq x \leq 6$. The graph of f' has horizontal tangents at $x = -2.5$, $x = -1.5$, and $x = 2$.

a) Find all values of x, $-4 \leq x \leq 6$, where f attains a relative maximum.

b) Find all values of x, $-4 \leq x \leq 6$, where f has inflection points. Justify your answers.

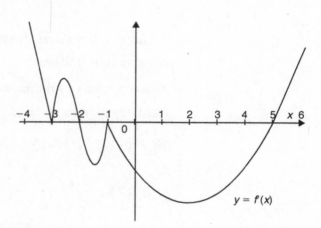

$y = f(x)$

SOLUTIONS: (a) f': $\underline{+++0+++0---0-----------0+++}$
$-4\quad -3\qquad -2\qquad -1\qquad\qquad\qquad\qquad 5\quad\ 6$

f has a relative maximum at $x = -2$ only because f' switches from positive to negative.

(b) $f'': \dfrac{---0++0-----0++0------0++++++++}{\quad -4 \quad -3 \; -2.5 \qquad -1.5 \; -1 \qquad\qquad 2 \qquad\qquad\qquad 6}$

f has inflection pts. at $x = -3, x = -2.5, x = -1.5, x = -1$,

and $x = 2$ because f'' switches signs at those values.

EXAMPLE 22:   Let f be a differentiable curve such that $f(2) = -1$. The figure below shows the graph of $y = f'(x)$, the derivative of the function f. The graph of f' has x-intercepts at $x = -1$ and $x = 3$ and a horizontal tangent at $x = 2$. Let g be the function $g(x) = e^{-f(x)}$.

a) Write the equation of the line tangent to g at $x = 2$.

b) Find all values of x where g has a relative minimum. Justify your answer.

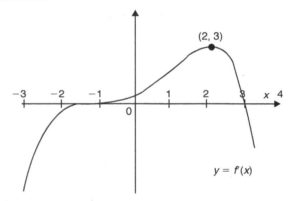

SOLUTIONS: (a) $g(2) = e^{-f(2)} = e^{-(-1)} = e$

$g'(x) = e^{-f(x)}(-f'(x))$

$g'(2) = e^{-f(2)}(-f'(2)) = e(-3) = -3e$

Tangent line: $y - e = -3e(x - 2)$

(b) Since $g'(x) = e^{-f(x)}(-f'(x))$ and $e^{-f(x)} > 0$, then g' changes from negative to positive when $f'(x)$ changes from positive to negative. This occurs at $x = 3$.

EXAMPLE 23:   The figure on the next page shows the graph of f', the derivative of the even function f. This graph has horizontal tangents at $x = 2$ and $x = 6$. The domain of f is $-8 \le x \le 8$ and $f(6) = -2$.

a) For what values of x on $[-8, 8]$ does f have a relative minimum? Justify your answer.

b) On what value of x is f concave up? Justify your answer.

c) Find the equation of tangent line to f at $x = -6$.

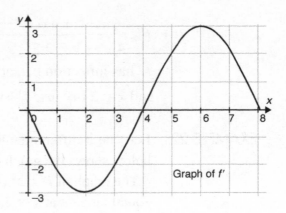

Graph of f'

SOLUTIONS: (a) Since f is even, f' will be odd.

$$f': 0 -----0+++++++0-------0+++++0$$
$$\quad -8 \qquad\quad -4 \qquad\qquad 0 \qquad\qquad 4 \qquad\quad 8$$

f has relative minima at $x = -4$ and $x = 4$ as f' switches from negative to positive.

(b) Since f is even, f'' will also be even.

$$f'': ----0++++++0----------0++++++0----$$
$$\quad -8 \quad -6 \qquad\quad -2 \qquad\qquad 2 \qquad\quad 6 \quad 8$$

f has inflection points at $x = -6, x = -2, x = 2, x = 6$ as f'' switches signs.

(c) Since f is even, $f(-6) = f(6) = -2$ and $f'(-6) = -f'(6) = -3$.

Tangent line : $y + 2 = -3(x + 6)$

TEST TIP

The AP exam is usually graded differently than the way your calculus teacher grades. For instance, it is not necessary to simplify equations in the free-response section of the AP exam. In fact, if you simplify an equation and make an error, you will lose points. So save time and do not simplify by multiplying out expressions or combining like terms. However, in the multiple-choice problems, you may have to simplify equations to match one of the answers.

EXAMPLE 24:   Let f be the function defined by $f(x) = k \sin x - e^{2x}$ where k is a constant. If f has a critical point at $x = \pi$, find the value of k and determine whether this point is a relative maximum, relative minimum or neither. Explain your reasoning.

SOLUTION: $f'(x) = k \cos x - 2e^{2x}$

$f'(\pi) = -k - 2e^{2\pi} = 0$ so $k = -2e^{2\pi}$.

$f''(x) = -k \sin x - 4e^{2x}$

$f''(\pi) = 0 - 4e^{2\pi} < 0$

By the 2nd derivative test, there is a relative maximum at $x = k$.

TEST TIP

A favorite type of problem on the AP exam asks students to use tangent lines to approximate values of functions at specific values. Example 25 uses techniques of function analysis to determine whether these are over- or under-approximations.

EXAMPLE 25:   Given $f(x) = x^3 - x^2 - 2x - 1$, as shown in the graph below, use the tangent line at $x = 1$ to approximate $f(0.9)$ and determine whether this is an over-approximation or under-approximation to $f(0.9)$.

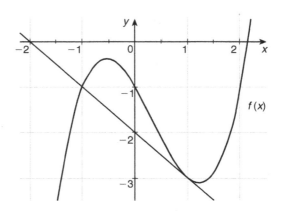

SOLUTION: $f(1) = 1 - 1 - 2 - 1 = -3$

$f'(x) = 3x^2 - 2x - 2$ and $f'(1) = 3 - 2 - 2 = -1$

Tangent line: $y + 3 = -1(x - 1)$ or $y = -x - 2$

$f(0.9) \approx -0.9 - 2 = -2.9$

$f''(x) = 6x - 2$ and $f''(1) = 6 - 2 = 4$

Since $f'(1) < 0$ and $f''(1) > 0$, f is decreasing and concave up at $x = 1$

So -2.9 must be an under-approximation to $f(0.9)$.

Absolute Extrema

Overview: Functions whose domain is all real numbers may have several relative extrema, but it is possible that they have none. An example is $f(x) = x^3$ which has a critical point at $x = 0$, but this critical point is neither a relative minimum nor a relative maximum. However, once we restrict the domain to a specific interval, a function *must* have an **absolute maximum** and an **absolute minimum**. The absolute maximum (sometimes called global maximum) is the largest value of $f(x)$ on an interval while the absolute minimum (sometimes called global minimum) is the smallest value of $f(x)$ on the interval. When you are given a function on a domain that is a specific interval and asked to find its range, you are being asked to find the difference between the absolute maximum value of the function and the smallest.

Even if the domain is all real numbers, it is possible that a function can have an absolute maximum or minimum or both. For instance, $f(x) = x^2$ graphs a parabola and we know that the absolute minimum value of the function is zero. $f(x) = \sin x$ graphs a wave and we know that the absolute maximum value of the function is 1 and the absolute minimum is -1, each occurring at an infinite number of points.

The general technique to find an absolute maximum or absolute minimum of $f(x)$ on the interval $[a, b]$ is:

(a) Find critical values of $f(x)$ (x-values where $f'(x) = 0$ or $f'(x)$ is not defined).

(b) Use the first derivative test to locate relative extrema of $f(x)$. (This step is optional).

(c) Evaluate $f(x)$ at critical values *and* at the endpoints. The largest of these values is the absolute maximum and the smallest of these is the absolute minimum.

EXAMPLE 26: Find the absolute minimum and maximum value of
$y = \left| x^2 - 5x - 14 \right| - x$ for $0 \le x \le 8$.

SOLUTION: $\left| x^2 - 5x - 14 \right| - x = \left| (x-7)(x+2) \right| - x$

$$y = \begin{cases} x^2 - 5x - 14 - x & \text{if } x \ge 7 \text{ or } x \le -2 \\ -x^2 + 5x + 14 - x & \text{if } -2 \le x \le 7 \end{cases}$$

$$y' = \begin{cases} 2x - 6 & \text{if } x \ge 7 \text{ or } x \le -2 \\ -2x + 4 & \text{if } -2 \le x \le 7 \end{cases}$$

$2x - 6 = 0 \Rightarrow x = 3$ but 3 is not in intervals $x \ge 7$ or $x \le -2$

$-2x + 4 = 0 \Rightarrow x = 2$

$y(0) = 14, \quad y(2) = 18, \quad y(7) = -7, \quad y(8) = 2.$

The absolute maximum value of y is 18.

The absolute minimum value of y is -7.

TEST TIP

Pay attention to the wording of absolute extrema problems on the AP exam. If you are asked for the absolute maximum value of a function on an interval, the value you want is the y-value where the absolute maximum occurs. The absolute minimum of $f(x) = x^2 + 3$ is 3. The absolute maximum occurs at $x = 0$. Confusing these concepts may cause you to make a mistake on a multiple-choice question you essentially understand.

EXAMPLE 27: (Calculator Active) Find the maximum acceleration of a particle moving along a straight line with velocity $v(t) = \ln(t + 1) + \cos 2t$ for $0 \le t \le \pi$. Show your reasoning.

SOLUTION: To maximize acceleration, first find acceleration, then take its derivative.

$$a(t) = \frac{1}{t+1} - 2\sin 2t$$

$$a'(t) = \frac{-1}{(t+1)^2} - 4\cos 2t = 0 \Rightarrow t = 0.823, t = 2.345$$

t	$a(t)$
0	1
0.823	−1.446
2.345	2.298
π	0.241

Maximum acceleration is 2.298 occuring at $t = 2.345$.

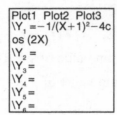 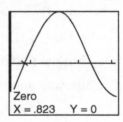 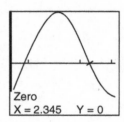

TEST TIP

On the calculator-active section of the AP exam, remember that while you may use the calculator's ZERO function to find roots of equations, using the calculator's MAXIMUM or MINIMUM function is not an allowable justification. You must use calculus to show your reasoning before you use the calculator to do the computation.

EXAMPLE 28: A particle moves along the x-axis whose position function is given by $x(t) = 2\sin\pi(t) - 3$ for $0 \le t \le \frac{5}{6}$. How far does the particle travel for $0 \le t \le \frac{5}{6}$?

SOLUTION: $v(t) = 2\pi \cos \pi t$

$v(t): \underset{0 \qquad\quad \frac{1}{2} \qquad\quad \frac{5}{6}}{+++++++0--------}$

The particle moves right from 0 to $\dfrac{1}{2}$ and then moves left.

At $t = 0$, the particle is at position $x(0) = -3$

At $t = \dfrac{1}{2}$, the particle is at position $x\left(\dfrac{1}{2}\right) = -1$

On $\left[0, \dfrac{1}{2}\right]$ the particle has moved from -3 to -1, a distance of 2 units.

At $t = \dfrac{5}{6}$, the particle is at position $x\left(\dfrac{5}{6}\right) = -2$

On $\left[\dfrac{1}{2}, \dfrac{5}{6}\right]$ the particle has moved from -1 to -2, a distance of 1 unit.

So on $\left[0, \dfrac{5}{6}\right]$ the particle has moved a distance of 3 units.

Note: Using integral calculus, the distance traveled would be expressed as $\displaystyle\int_0^{5/6} |2\pi \cos \pi t|\, dt$.

Knowing $x(t)$ allows students to solve this problem with differential calculus.

Optimization

Overview: To optimize we are interested in finding the largest or smallest value of some quantity that fulfills a specific requirement. You may be interested in a route that gets you from home to school with the *shortest distance* if you are walking or the *shortest time* if you are driving. These routes might not be the same. You might want to buy a car with the *most options* on it with a set amount of money or you might want to buy a car with specific options for the *least amount of money*. These are examples of optimization problems. And while optimization problems that are presented as word problems do not occur frequently on the AP exam, they are important enough to discuss.

To solve an optimization problem, you must recognize it. While related rates problems can be identified with words relating to change (increasing, decreasing, etc), optimization problems' telltale signs are phrases like "minimize area," "greatest volume," and "cheapest price."

Procedure for Solving an Optimization Problem

(1) Assign variables to all given quantities and quantities you wish to find. If it helps, make a sketch of the problem situation using those variables.

(2) Write a "primary equation" for the variable that needs to be maximized or minimized. Typical equations are:

Area = length · width   Volume = length · width · height

Volume = $\pi r^2 h$ Hypotenuse = $\sqrt{a^2 + b^2}$

(3) Reduce the right side of this primary equation to one having a single variable. If there is more than one unknown variable on the right side, you must write a "secondary equation" (a restriction or a constraint) relating the variables of the primary equation.

(4) After applying the secondary equation to the primary equation, take the derivative of the primary equation and set it equal to zero. Make sure that these solutions make sense in the problem situation. For instance, time is rarely negative and you cannot have a length of 20 feet when you only have 15 feet of fencing.

(5) Plug these solutions and endpoints into the primary equation and determine which gives the maximum or minimum value. If there is a specific interval given, endpoints must be tested as well.

(6) Make sure you answer the question that is asked. If you are asked to find the maximum profit you can make when you are burning CD's, the answer will be monetary even though your equation will give you how many CD's to burn to generate this maximum profit.

EXAMPLE 29: A rectangle is inscribed bounded by the graph of $y = 8 - x^3$ in the first quadrant, the x-axis and the y-axis as shown in the figure below. Find the largest area rectangle that can be inscribed.

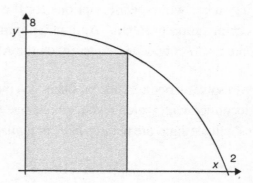

SOLUTION: Primary: Area $= x \cdot y$ Secondary: $y = 8 - x^3$

$$A = x(8 - x^3) = 8x - x^4$$
$$0 = 8 - 4x^3$$
$$4x^3 = 8 \text{ so } x = \sqrt[3]{2}$$
$$\text{Maximum Area} = \sqrt[3]{2}(8-2) = 6\sqrt[3]{2}$$

TEST TIP

When optimization questions are asked, be aware that your answer could generate a maximum or a minimum. Although no work is shown to justify that the solution in Example 29 is the largest area rectangle, it should be apparent that values of x closer to 2 would give smaller and smaller rectangles. In optimization questions on the AP exam, typically the solution from the technique shown is the answer to the problem and students do not have to justify that it is indeed the maximum unless they are specifically asked to justify the solution.

EXAMPLE 30: (Calculator Active) A cost of a gallon of milk changes dramatically over a 7-day period. The price of a gallon of milk is modeled by the equation $P(t) = \sqrt{2t + 2} + e^{-t}$ where t is measured in days. On what day is the price of milk changing the fastest? Express this change using correct units.

SOLUTION: We want to maximize $P'(t)$, not $P(t)$, so we first have to find $P'(t)$, then take its derivative.

$$P'(t) = \frac{1}{2}(2t + 2)^{-1/2}(2) - e^{-t}$$

$$P''(t) = \frac{-1}{2}(2t + 2)^{-3/2}(2) + e^{-t} = \frac{-1}{(2t + 2)^{3/2}} + e^{-t} = 0$$

t	$P'(t)$
0	-0.293
3.188	0.304
7	0.249

The price of milk is increasing the fastest when $t = 3.188$. At that time, the price is increasing at the rate of 30.4 cents per day. Note that on day 0, the price is decreasing at the rate of 29.3 cents per day.

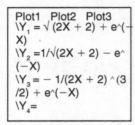

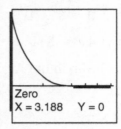

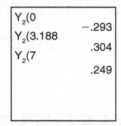

TEST TIP

If you are asked to find the time when a quantity is changing the fastest, you must first write an expression that represents that change. You then maximize it by taking *its* derivative. In Example 30, you are asked to maximize the change of the price, so you must first write an expression representing the change of price. If this were a multiple-choice question on the AP exam, one of the choices would be based on setting $P'(t) = 0$ as opposed to the correct solution of setting $P''(t) = 0$.

Time for a quiz
• Review strategies in Chapter 2
• Take Quiz 4 at the REA Study Center

(www.rea.com/studycenter)

Take Mini-Test 1
on Chapters 3–6
Go to the REA Study Center
(www.rea.com/studycenter)

Integration

Antidifferentiation and Indefinite Integrals

Overview: The first part of the course concerned itself with taking derivatives. As we do with most mathematical concepts, we learn a concept and then we learn the inverse of that process. Addition, then subtraction; multiplication, then division; squares, then square roots; exponentiation, then logs; that is how math is taught. So it is natural that the next step in the process following differentiation will be **antidifferentiation**.

We know that the derivative of x^2 is $2x$, so it could be said that an **antiderivative** of $2x$ is x^2. However, the derivative of x^2, $x^2 + 5$, $x^2 - 12$, $x^2 + \pi$ are all $2x$ and if we reverse the process, we won't know the particular expression to go back to. So we write the antiderivative of $2x$ as $x^2 + C$ where C is a constant.

This can be written mathematically as: $\int 2x \, dx = x^2 + C$. The $\int$ symbol is called an integral sign and this process is called **indefinite integration**. The C is called the constant of integration and must always be present when performing indefinite integration. The dx has meaning that is explained in the next section but, for now, it represents the variable of importance when integrating.

While a derivative can be taken for any expression, not all expressions can be integrated. Just as differentiation of complicated expressions involves u-substitution, substitution must be used to integrate some expressions. It is important to include the constant of integration when performing anti-differentiation.

Students need to know these integration formulas. Some appear on the AP exam more than others.

Most likely to appear on the AP exam

$$\int u^n\, du = \frac{u^{n+1}}{n+1} + C,\, n \neq -1$$

$$\int \frac{du}{u} = \ln|u| + C$$

$$\int e^u\, du = e^u + C$$

$$\int \sin u\ du = -\cos u + C$$

$$\int \cos u\ du = \sin u + C$$

$$\int \sec^2 u\ du = \tan u + C$$

$$\int \frac{du}{a^2 + u^2} = \frac{1}{a} \tan^{-1} \frac{u}{a} + C$$

Less likely to appear on the AP exam

$$\int \csc^2 u\ du = -\cot u + C$$

$$\int \sec u \tan u\ du = \sec u + C$$

$$\int \csc u \cot u\ du = -\csc u + C$$

$$\int a^u\ du = \frac{1}{\ln a} a^u + C$$

$$\int \frac{du}{\sqrt{a^2 - u^2}} = \sin^{-1} \frac{u}{a} + C$$

There are also some basic rules for all integrals that are vital to know:

$$\int [f(x) \pm g(x)]\ dx = \int f(x)\ dx \pm \int g(x)\ dx + C$$

$$\int k f(x)\ dx = k \int f(x)\ dx \text{ where } k \text{ is a constant}$$

EXAMPLE 1: Find $\int \left(x^3 + 4x^2 - 3x + 1 \right) dx.$

SOLUTION: This uses the power rule for all terms:

$$\int \left(x^3 + 4x^2 - 3x + 1 \right) dx = \frac{x^4}{4} + \frac{4x^3}{3} - \frac{3x^2}{2} + x + C.$$

EXAMPLE 2: Find $\int \left(2x^2 + 3 \right)^2 dx.$

SOLUTION: This expression could be differentiated using u-substitution. For this integration problem, u-substitution is not an option. The only way to integrate is to multiply out and then apply the power rule:

$$\int \left(2x^2 + 3 \right)^2 dx = \int \left(4x^4 + 12x^2 + 9 \right) dx = \frac{4x^5}{5} + 4x^3 + 9x + C$$

EXAMPLE 3: Find $\int \left(4x - 9 \right)^5 dx.$

SOLUTION: While it is possible to do this problem by multiplying out and then integrating with the power rule, it is much easier to use u-substitution. Typically in problems like this, we let u be the expression in the parentheses.

$u = 4x - 9$ so $\dfrac{du}{dx} = 4$ and $du = 4dx$

We need to put $\int \left(4x - 9 \right)^5 dx$ in terms of u. To get a 4 dx into the problem, we need to multiply the dx by 4 so we compensate by multiplying the entire expression by $\dfrac{1}{4}$.

$$\int \left(4x - 9 \right)^5 dx = \frac{1}{4} \int \left(4x - 9 \right)^5 (4\ dx) = \frac{1}{4} \int u^5 du.$$

We now integrate in terms of u: $\dfrac{1}{4} \int u^5 du = \dfrac{1}{4} \left(\dfrac{u^6}{6} \right)$

And finally change back to the variable x and add the constant of integration: $\dfrac{1}{24} \left(4x - 9 \right)^6 + C.$

EXAMPLE 4: Find $\int \dfrac{x}{\sqrt{x^2+4}}\,dx$.

SOLUTION: $\displaystyle\int x\left(x^2+4\right)^{-1/2}\,dx$ $u = x^2 + 4$
$du = 2x\,dx$

$\displaystyle = \frac{1}{2}\int 2x\left(x^2+4\right)^{-1/2}\,dx$

$\displaystyle = \frac{1}{2}\int u^{-1/2}\,du$

$\displaystyle = \frac{1}{2}\left(\frac{u^{1/2}}{\frac{1}{2}}\right) = \frac{1}{2}u^{1/2}(2)$

$\displaystyle = \sqrt{x^2+4}+C$

EXAMPLE 5: Find $\int \dfrac{\sin\left(\dfrac{1}{x}\right)}{x^2}\,dx$.

SOLUTION: $\displaystyle -\int \frac{\sin\left(\dfrac{1}{x}\right)}{x^2}(-dx)$ $u = \dfrac{1}{x}$

$\displaystyle = -\int \sin u\,du$ $du = \dfrac{-1}{x^2}\,dx$

$\displaystyle = -(-\cos u) = \cos\left(\frac{1}{x}\right)+C$

TEST TIP

Students find it easy to confuse the derivative rules and integration rules for sine and cosine. Considering that they are so vital on the AP exam, it is important to get them right. Here is a good way to memorize them: The derivative of the trig function is the one below it. The integral of the trig function is the one above it.

$$\sin x$$
$$\downarrow \quad \uparrow$$
$$\cos x$$
$$\downarrow \quad \uparrow$$

Derivative $\qquad -\sin x \qquad$ Integration

Process $\qquad\qquad\qquad$ Process
$$\downarrow \quad \uparrow$$
$$-\cos x$$
$$\downarrow \quad \uparrow$$
$$\sin x$$

EXAMPLE 6: Find $\int \dfrac{x^2}{x^3+1}\,dx$.

SOLUTION: When integrating a fraction with variables in both the numerator and the denominator, a technique that usually (but not always) works is to let the u be the expression in the denominator. If the numerator can be written as a constant factor of the denominator, you can use the formula: $\int \dfrac{du}{u} = \ln|u| + C$.

$$\int \frac{x^2}{x^3+1}\,dx \qquad\qquad u = x^3 + 1$$

$$\qquad\qquad\qquad\qquad du = 3x^2\,dx$$

$$= \frac{1}{3}\int \frac{x^2}{x^3+1}\,(3dx)$$

$$= \frac{1}{3}\int \frac{du}{u} = \frac{1}{3}\ln\left|x^3+1\right| + C$$

EXAMPLE 7: Find $\int \tan(1-5x)\,dx$.

SOLUTION: The integral of the tangent function does not appear in the basic list of expressions that are easily integrated. But by expressing the tangent function as $\dfrac{\sin\theta}{\cos\theta}$, this problem can be done.

$$\int \tan(1-5x)\,dx \qquad\qquad u = \cos(1-5x)$$

$$= \int \frac{\sin(1-5x)}{\cos(1-5x)}\,dx \qquad du = -5\sin(1-5x)(-5)\,dx$$

$$= 5\sin(1-5x)\,dx$$

$$= \frac{1}{5}\int \frac{\sin(1-5x)}{\cos(1-5x)}(5dx)$$

$$= \frac{1}{5}\int \frac{du}{u} = \frac{1}{5}\ln|\cos(1-5x)| + C$$

EXAMPLE 8: Find $\int \dfrac{\ln 2x}{x}\,dx$.

SOLUTION: The technique of letting the u represent the denominator clearly doesn't work here. So, let the u be the expression in the numerator. Good things happen!

$$\int \frac{\ln 2x}{x}\,dx \qquad\qquad u = \ln 2x$$

$$= \int u\,du \qquad du = \frac{1}{2x}(2dx) = \frac{1}{x}dx$$

$$= \frac{u^2}{2} = \frac{(\ln 2x)^2}{2} + C$$

EXAMPLE 9: Find $\int xe^{1-5x^2}\,dx$.

SOLUTION: Many times when you are integrating an expression with e to a power, the integration will work if you let u be the power.

$$\int xe^{1-5x^2}\,dx \qquad\qquad u = 1-5x^2$$

$$= \frac{-1}{10}\int -10xe^{1-5x^2}\,dx \qquad du = -10x\,dx$$

$$= \frac{-1}{10}\int e^u\,du = \frac{-1}{10}e^{1-5x^2} + C$$

EXAMPLE 10: Find $\int \dfrac{1}{x^2 + 25}\, dx.$

SOLUTION: Letting the denominator represent the u clearly doesn't work here. The only way to do this is to recognize its form from the integration rules above.

$$\int \frac{1}{x^2 + 25}\, dx \qquad\qquad u = x,\ a = 5$$
$$= \int \frac{1}{u^2 + a^2}\, du \qquad\qquad du = dx$$
$$= \frac{1}{a} \tan^{-1}\left(\frac{u}{a}\right) = \frac{1}{5} \tan^{-1}\left(\frac{x}{5}\right) + C$$

TEST TIP

Integration is practice, practice, practice and many times, trial and error. Hopefully your calculus course has done a lot of drilling on integration techniques. You will certainly be tested on it on the AP exam. If there is a tough problem that appears in the multiple-choice section of the exam that you cannot do, you can always take the derivative of each of your 5 answers and see which one is the original expression. It is potentially time-consuming, but it works!

DIDYOUKNOW?

An electrical engineer uses integration to determine the exact length of electrical wire needed to connect two substations that are miles apart. If the wire were in straight lines, basic arithmetic could be used. But because the cable is hung from poles, it is constantly curving in the shape of a catenary. The use of calculus allows a very precise figure to be calculated.

Area and the Definite Integral

Overview: The second part of a typical calculus course is called **integral calculus**. Differential calculus concerned itself with the problem of finding the slope of a tangent line to a curve at a point. Armed with that information, using linear approximation, we could approximate a value of $f(c)$ at point P close to the point of tangency.

In integral calculus we are concerned with finding the area between a curve, the x-axis and lines $x = a$ and $x = b$. Just as the slope of the tangent line had a mathematical name, the derivative, the area between a function $f(x)$ and the x-axis between a and b is called a **definite integral** and is written as $\int_a^b f(x)\, dx$. This is different from an indefinite integral $\int f(x)\, dx$ as studied in the previous section. The connection between the indefinite and definite integral will be studied later in this chapter when we look at the Fundamental Theorem of Calculus.

The notation makes sense by looking at the left-hand figure below. We draw a rectangle whose height is $f(x)$, the height of the rectangle above the x-axis. We define dx as the width of the rectangle. So $f(x)\, dx$ represents (height) • (width) = the area of any one rectangle. The definite integral sign $\int_a^b$ represents the sum of the areas of infinitely thin rectangles as shown in the right-hand figure below. The a represents the starting place for these rectangles while the b represents the ending place for these rectangles. Remember that definite integrals do not have a $+ C$ when calculated because the area is a specific number. That is why they are called *definite* integrals.

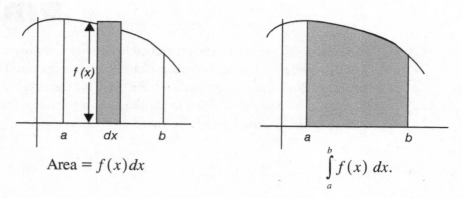

$$\text{Area} = f(x)\,dx \qquad\qquad \int_a^b f(x)\, dx.$$

When you see a definite integral, you must always think of it as a representation of an area. In geometry you learned that area is always a positive number. However, it is possible for a definite integral to be negative. Here's how:

When $a < b$, we determine the area under the curve from left to right. Our dx will be a positive number.

When $b < a$, we determine the area under the curve from right to left. Our dx will be a negative number.

This can be summarized below.

	$f(x) > 0$ (curve above x-axis)	$f(x) < 0$ (curve below x-axis)
$a < b$ ($dx > 0$) (left to right)	$\int_a^b f(x)\ dx > 0$	$\int_a^b f(x)\ dx < 0$
$b < a$ ($dx < 0$) (right to left)	$\int_a^b f(x)\ dx < 0$	$\int_a^b f(x)\ dx > 0$

Furthermore, there are several rules that make absolute sense when you think of definite integrals as areas:

1. $\int_a^a f(x)\ dx = 0$ — If you start at a and end at a, there is no area.

2. $\int_b^a f(x)\ dx = -\int_a^b f(x)\ dx$ — Suppose $f(x)$ is positive. $\int_a^b f(x)\ dx$ will be positive so $-\int_a^b f(x)\ dx$ will be negative which will be $\int_b^a f(x)\ dx$. You can always change the order of the limits in a definite integral if you apply a negative sign to the integral sign.

3. $\int_a^b f(x)\ dx + \int_b^c f(x)\ dx = \int_a^c f(x)\ dx$ — Add the area from a to b to the area from b to c and you get the area from a to c. It makes no difference if the function is above or below the x-axis.

4. $\int_a^b kf(x)\ dx = k\int_a^b f(x)\ dx$ where k is a constant — The area under $k \cdot f(x)$ is k times the area under $f(x)$.

5. $\int_a^b [f(x) \pm g(x)]\ dx = \int_a^b f(x)\ dx \pm \int_a^b g(x)\ dx$ — The area under $f(x)$ plus (or minus) $g(x)$ is the area under $f(x) + g(x)$ or $f(x) - g(x)$.

EXAMPLE 11:   For the following problems suppose $f(x)$ and $g(x)$ are given by the following graphs, made up of lines and a semi-circle. Evaluate each part.

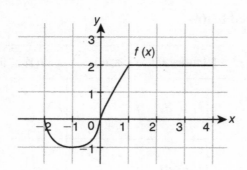   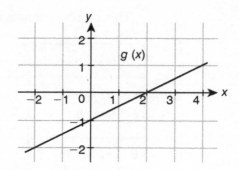

a) $\int\limits_{0}^{1} f(x)\, dx$

b) $\int\limits_{1}^{4} f(x)\, dx$

c) $\int\limits_{0}^{4} f(x)\, dx$

d) $\int\limits_{4}^{0} f(x)\, dx$

e) $\int\limits_{-2}^{0} f(x)\, dx$

f) $\int\limits_{0}^{-2} f(x)\, dx$

g) $\int\limits_{-2}^{4} 4f(x)\, dx$

h) $\int\limits_{-2}^{4} g(x)\, dx$

i) $\int\limits_{-2}^{4} [f(x) - g(x)]\, dx$

j) $\left| \int\limits_{-2}^{4} g(x)\, dx \right|$

k) $\int\limits_{-2}^{4} |g(x)|\, dx$

l) $\int\limits_{-2}^{4} [f(x) + 3]\, dx$

SOLUTIONS: a)   This is the area of a triangle with base 1 and height 2. Answer is $\frac{1}{2}(1)(2) = 1$.

b)   This is the area of a rectangle with base 3 and height 2. Answer is $3(2) = 6$.

c)   This utilizes rule 3 above. Answer is $1 + 6 = 7$.

d)   This utilizes rule 2 above. If $\int\limits_{0}^{4} f(x)\, dx = 7$, then $\int\limits_{4}^{0} f(x)\, dx = -7$.

e) This is the area of a semicircle. The dx is positive (left to right) but the function is negative (below the axis). The answer is

$$= -\frac{1}{2}\pi(1)^2 = -\frac{\pi}{2}.$$

f) Rule 2 again. Or the function is negative and the dx is negative (right to left) so the answer is $\frac{\pi}{2}$.

g) A combination of rules 3 and 4.

$$\int_{-2}^{4} f(x)\,dx = \int_{-2}^{0} f(x)\,dx + \int_{0}^{4} f(x)\,dx = -\frac{\pi}{2}+7. \text{ So}$$

$$\int_{-2}^{4} 4f(x)\,dx = 4\int_{-2}^{4} f(x)\,dx = 4\left(-\frac{\pi}{2}+7\right) = -2\pi+28.$$

h) Using rule 3, we have two triangles.

$$\int_{-2}^{4} g(x)\,dx = \int_{-2}^{2} g(x)\,dx + \int_{2}^{4} g(x)\,dx = \frac{1}{2}(4)(-2)+\frac{1}{2}(2)(1) = -3.$$

i) Rule 5:

$$\int_{-2}^{4} [f(x)-g(x)]\,dx = \int_{-2}^{4} f(x)\,dx - \int_{-2}^{4} g(x)\,dx = -\frac{\pi}{2}+7-(-3) = 10-\frac{\pi}{2}.$$

j) $\left| \int_{-2}^{4} g(x)\,dx \right| = |-3| = 3.$

k) The absolute value of the g function takes anything that is negative (below the x-axis) and makes it positive. The g function is negative between -2 and 2, so that integral must become positive.

$$\int_{-2}^{4} |g(x)|\,dx = -\left(\int_{-2}^{2} g(x)\,dx \right) + \int_{2}^{4} g(x)\,dx = -\left[\frac{1}{2}(4)(-2) \right] + \frac{1}{2}(2)(1) = 5.$$

Note the subtle but important difference between problem j) and problem k).

l) Rule 5: $\int_{-2}^{4} [f(x)+3] \, dx = \int_{-2}^{4} f(x) \, dx + \int_{-2}^{4} 3 \, dx.$ $\int_{-2}^{4} 3 \, dx$ means the area under the function $y = 3$ (which is a horizontal line) from -2 to 4. This is a rectangle with base 6 and height 3 and the area is 18. $\int_{-2}^{4} f(x) \, dx + \int_{-2}^{4} 3 \, dx = -\dfrac{\pi}{2} + 7 + 18 = 25 - \dfrac{\pi}{2}.$

EXAMPLE 12: Given the graph of $f(x)$ below with areas between the x-axis and f as shown, find the following:

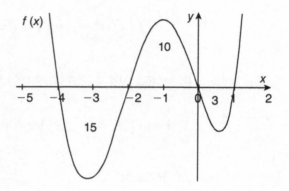

a) $\int_{-4}^{1} f(x) \, dx$ b) $\left| \int_{-4}^{1} f(x) \, dx \right|$ c) $\int_{-4}^{1} |f(x)| \, dx$

SOLUTIONS: a) $\int_{-4}^{1} f(x) \, dx = -15 + 10 - 3 = -8$

b) $\left| \int_{-4}^{1} f(x) \, dx \right| = |-8| = 8$

c) $\int_{-4}^{1} |f(x)| \, dx = -(-15) + 10 - (-3) = 28$

DIDYOU**KNOW?**

The integral symbol $\int$ was introduced by the German Gottfried Wilhelm von Leibniz at the end of the 17th century. The symbol was used because it resembled the letter S standing for an infinite Sum.

The Accumulation Function

Overview: Definite integrals refer to the area under a curve. The limits of integration indicate the x-values where the area process starts and the area process ends. We now make the upper limit of integration a variable, usually x. Our integral structure is now changed to: $g(x) = \int_{a}^{x} f(t)\, dt$. This says we will start finding the area process at some constant a and end the area process at some variable x. Since this is a function of the variable x, we write $f(t)\, dt$ rather than $f(x)\, dx$ to prevent confusion. This is sometimes called the **accumulation function** because we are accumulating area under a curve based on the value of x.

EXAMPLE 13:   Below is the graph of $y = f(x)$ on the domain $[-4, 5]$, made up of lines. Let $g(x) = \int_{0}^{x} f(t)\, dt$. Complete the chart and determine the maximum and minimum value of g on $[-4, 5]$.

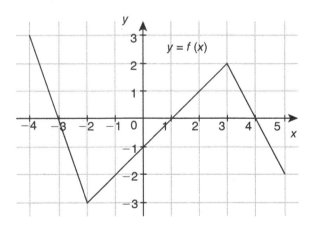

x	-4	-3	-2	-1	0	1	2	3	4	5
$g(x) = \int_0^x f(t)\, dt$										

SOLUTION: Start the chart in the middle: At $x = 0$, $g(0) = \int_0^0 f(t)\, dt$ which is the area under the f curve given above starting at 0 and ending at 0. We know this to be 0.

At $x = 1$, $g(1) = \int_0^1 f(t)\, dt$ which is the area of a triangle with f negative and dx positive. $g(1) = \frac{1}{2}(-1)(1) = -0.5$.

At $x = 2$, $g(2) = \int_0^2 f(t)\, dt$. We use our definite integration rules to say that $\int_0^2 f(t)\, dt = \int_0^1 f(t)\, dt + \int_1^2 f(t)\, dt$. This is why we call this structure the accumulation function. $g(2) = -0.5 + 0.5 = 0$ as the two triangular areas cancel each other out.

At $x = -1$, $g(-1) = \int_0^{-1} f(t)\, dt$. This is the area of the trapezoid between the function f and $x = 0$ and $x = -1$. You might prefer to view it as a square and a triangle. Realize that the function f is negative (below the x-axis) and the dt is negative as well as we are going from 0 to -1. So the integral is positive and the value is 1.5.

The rest of the chart is below.

x	-4	-3	-2	-1	0	1	2	3	4	5
$g(x) = \int_0^x f(t)\, dt$	4	5.5	4	1.5	0	-0.5	0	1.5	2.5	1.5

The maximum value of $g(x)$ is 5.5 at $x = -3$ and the minimum value of $g(x)$ is -0.5 at $x = 1$. Later in this chapter when we tackle the Second Fundamental Theorem of Calculus, we will see how to determine the maximum and minimum value of the accumulation function without filling out this cumbersome chart.

EXAMPLE 14: A graph of f is made up of lines and a semi-circle as shown below.
Let $F(x) = \int_{-3}^{x} f(t)\, dt$.

Find

a) $F(5)$

b) $F(-5)$.

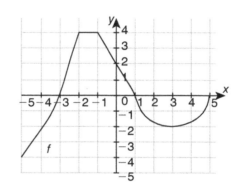

SOLUTIONS: The lower limit of the accumulation function does not have to be zero. It can start anywhere.

a) $F(5) = \int_{-3}^{5} f(t)\, dt = \int_{-3}^{1} f(t)\, dt + \int_{1}^{5} f(t)\, dt$. This is the area of a trapezoid and a semi-circle. The dx is positive so
$$F(5) = \int_{-3}^{5} f(t)\, dt = \frac{1}{2}(4)(4+1) - \frac{1}{2}\pi(2)^2 = 10 - 2\pi.$$

b) $F(-5) = \int_{-3}^{-5} f(t)\, dt$. This is the area of a triangle. The dx is negative and the function is negative, so the integral is positive.
$$F(-5) = \frac{1}{2}(2)(4) = 4.$$

EXAMPLE 15: Graph $y = \int_{1}^{x} (2t) \; dt$.

SOLUTION: $y = t^2 \Big|_{1}^{x} = x^2 - 1$ which we know to be a parabola.

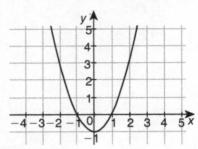

Note: It is possible to graph accumulation functions on the TI-84 calculator. Here is how they would be input:

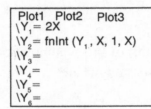
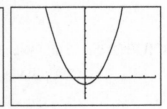

However, the graphing of this function is very slow as the calculator must compute the value of the definite integral at every x-value in the window.

Riemann Sums and the Trapezoidal Rule

Overview: Riemann sums are approximations for definite integrals, which we know represent areas under curves. There are numerous real-life models for areas under curves so this is an important concept. Typically, Riemann sum problems, named after Bernhard Riemann (1826–1866) who was the first to use them, show up when we are given data points as opposed to algebraic functions.

There are three types of Riemann sums to approximate the area under a curve: Left Riemann Sums, Right Riemann Sums, and Midpoint Riemann Sums. These use rectangles between the function and the x-axis. There is a fourth method called the Trapezoidal Rule, which uses trapezoids. The examples below use 4 rectangles, and in the last case, 4 trapezoids. Each of them provides an approximation to $\int_{a}^{b} f(x) \; dx$.

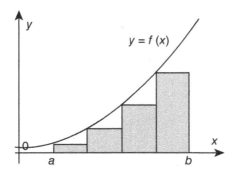

Left Riemann sums – the rectangles are built starting on the left side of the function. In this specific case, the Riemann sum will under-approximate the true area: $\int_a^b f(x)\,dx$

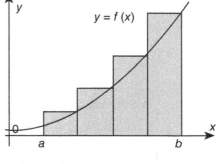

Right Riemann sums – the rectangles are built ending on the right side of the function. In this specific case, the Riemann sum will over-approximate the true area: $\int_a^b f(x)\,dx$

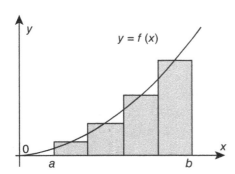

Midpoint Riemann sums – the rectangles are built halfway between the position of the left and right Riemann sums. This gives a very good approximation to the true area: $\int_a^b f(x)\,dx$

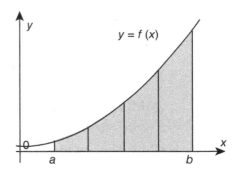

Trapezoids – 4 trapezoids are built. This is the average of the left and right Riemann sums and gives an excellent approximation to the actual area: $\int_a^b f(x)\,dx$

Given data points:

x	x_0	x_1	x_2	...	x_{n-2}	x_{n-1}	x_n
f(x)	$f(x_0)$	$f(x_1)$	$f(x_2)$	...	$f(x_{n-2})$	$f(x_{n-1})$	$f(x_n)$

Assuming equally spaced x-values: $x_{i+1} - x_i = b$ (which is the base of each rectangle)

Left Riemann Sums: $\int_{x_0}^{x_n} f(x)\, dx \approx b\left[f(x_0) + f(x_1) + f(x_2) + \ldots + f(x_{n-1}) \right]$

Right Riemann Sums: $\int_{x_0}^{x_n} f(x)\, dx \approx b\left[f(x_1) + f(x_2) + f(x_3) + \ldots + f(x_n) \right]$

Trapezoids: the average of the left and right Riemann sums: $\int_{x_0}^{x_n} f(x)\, dx \approx \dfrac{LRS + RRS}{2}$

or $\int_{x_0}^{x_n} f(x)\, dx \approx \dfrac{b}{2}\left[f(x_0) + 2f(x_1) + 2f(x_2) + \ldots + 2f(x_{n-2}) + 2f(x_{n-1}) + f(x_n) \right]$

If bases are not the same (typical in AP questions), you have to compute the area of each trapezoid: $\dfrac{1}{2}(x_{i+1} - x_i)\left[f(x_i) + f(x_{i+1}) \right]$ or $\dfrac{1}{2}b\left[f(x_i) + f(x_{i+1}) \right]$

Midpoints: This is commonly misunderstood. For example, you cannot draw a rectangle halfway between $x=2$ and $x=3$ because you may not know $f(2.5)$. You can't make up data. So in the table above the first midpoint rectangle would be drawn halfway between x_0 and x_2 which is x_1. So assuming that the x-values are equally spaced, the midpoint sum is $S = 2b\left[f(x_1) + f(x_3) + f(x_5) + \ldots f(x_{n-1}) \right]$.

EXAMPLE 16: Using the data points below, estimate $\int_{-1}^{17} f(x)\, dx$ using

a) 6 left Riemann sums
b) 6 right Riemann sums
c) 3 midpoint rectangles
d) 6 trapezoids.

x	-1	2	5	8	11	14	17
f(x)	6	2	7	4	-2	-5	3

SOLUTIONS: a) $\int_{-1}^{17} f(x)\,dx \approx 3(6+2+7+4-2-5)=36$

b) $\int_{-1}^{17} f(x)\,dx \approx 3(2+7+4-2-5+3)=27$

c) $\int_{-1}^{17} f(x)\,dx \approx 6(2+4-5)=6$. With only 3 rectangles, it is difficult to be accurate.

d) $\int_{-1}^{17} f(x)\,dx \approx \dfrac{36+27}{2}=31.5$ or

$\int_{-1}^{17} f(x)\,dx \approx \dfrac{1}{2}(3)\big[6+2(2)+2(7)+2(4)+2(-2)+2(-5)+3\big]=31.5$

Note that these are all approximations to the area under the curve and we have no idea how good these approximations are. The x-increment from point to point is 3 and we have no idea what kind of fluctuations occur between points. The only way to get more accuracy is to have more points. In calculus, we find the sum of an infinite number of very thin rectangles to actually find the area. This uses limits and is explored in the next section.

Finding area has greater importance when the area has significance in a word problem. If the x-axis is time measured in t and the function $v(t)$ is velocity measured in $\dfrac{\text{distance}}{\text{time}}$, the area of any one rectangle is measured as $\text{time}\left(\dfrac{\text{distance}}{\text{time}}\right)$ and thus $\int_{t_1}^{t_2} v(t)\,dt$ represents the distance a particle travels between t_1 and t_2 if the particle only travels in one direction.

EXAMPLE 17:   (Calculator Active) A fish is traveling along a straight line. During the time interval $0 \le t \le 45$ seconds, its velocity v, measured in ft/sec is a continuous function. The table below shows the fish's velocity at selected times. Using the trapezoidal rule, approximate $\dfrac{1}{45}\int_{0}^{45}|v(t)|\,dt$ and interpret its meaning in the context of the problem.

t (sec)	0	10	15	30	45
$v(t)$ (ft/sec)	-8	2	12	7	-3

SOLUTION: The first issue is the fact that we want the absolute value of v, which means the speed of the fish. So all negative velocities must be positive. The second issue is that the time increments are not uniform so each trapezoid must be found individually.

$$\int_0^{45} |v(t)|\, dt = \frac{1}{2}(10)(8+2) + \frac{1}{2}(5)(2+12) + \frac{1}{2}(15)(12+7) + \frac{1}{2}(15)(7+3) = 302.5$$

This will be measured in feet and represents the distance the fish traveled in the 45 seconds. So $\frac{1}{45}\int_0^{45} |v(t)|\, dt = 6.722 \dfrac{\text{ft}}{\text{sec}}$ represents the average velocity (or average speed) of the fish over the 45 seconds.

EXAMPLE 18: Ted is exercising on a stair-stepper machine. The rate of climbing in steps/minute is given by a differentiable function r of time t. The table below shows his rate every 2 minutes over a 20-minute period.

a) Find the difference in the estimation of $\int_0^{20} r(t)\, dt$ using left Riemann sums and right Riemann sums.

b) If 22 steps represent climbing a floor of a building, use the trapezoidal rule to approximate the number of floors (nearest integer) Ted climbed in 20 minutes.

t (minutes)	$r(t)$ steps per minutes
0	5
2	70
4	80
6	85
8	90
10	95
12	90
14	90
16	80
18	70
20	80

SOLUTIONS: a) Left: $\int_0^{20} r(t)\ dt \approx 2(5+70+80...+70)=1510$

Right: $\int_0^{20} r(t)\ dt \approx 2(70+80+85+...+80)=1660$

Difference: $1660-1510=150$

Note: These calculations have the same numbers except for the first and last, so the difference $=2(80-5)=150$.

b)

Since you have computed LRS and RRS, by the Trapezoidal Rule

$$\int_0^{20} r(t)\ dt \approx \frac{1510+1660}{2}=1585.$$

or $\int_0^{20} r(t)\ dt \approx \frac{2}{2}(5+2(70)+2(80)+2(85)+...+80)=1585$ steps.

$\frac{1585}{22}\approx 72$ floors.

EXAMPLE 19: Let $F(x)=\int_{-1}^{x} e^{\cos t}\ dt$. Use the Trapezoidal Rule with four equal subdivisions to approximate $F(1)$.

SOLUTION: base $=\dfrac{1-(-1)}{4}=\dfrac{1}{2}$

$$F(1)=\int_{-1}^{1} e^{\cos t}\ dt \approx \frac{1}{2}\left(\frac{1}{2}\right)\left[e^{\cos(-1)}+2e^{\cos(-0.5)}+2e^{\cos(0)}+2e^{\cos(0.5)}+e^{\cos(1)}\right]=\frac{18.490}{4}=4.622$$

DIDYOUKNOW?

There are other rules for approximating areas other than Riemann sums or the Trapezoidal Rule. Simpson's Rule (Thomas Simpson, 1710–1761) uses parabolas rather than trapezoids and because of the curved nature of parabolas, fewer are needed. Simpson's Rule is widely used by Naval architects to approximate cross-sectional areas and volumes of ships or lifeboats.

The Fundamental Theorem of Calculus

Overview: The **Fundamental Theorem of Calculus (FTC)** is the link between finding areas under curves and computing the indefinite integral of a function. Students should know that the FTC says that the **definite integral** $\int_a^b f(x)\ dx = F(b) - F(a)$ where $F(x)$ is an antiderivative of $f(x)$. This leads to the fact that $F(b) = F(a) + \int_a^b f(x)\ dx$. Never forget that when you see $\int_a^b f(x)\ dx$, you should always think of area under a curve. But the FTC allows us to calculate this area, especially when the function does not allow us to compute this area by geometric means.

EXAMPLE 20:   Find $\int_{-4}^{3} \left(12 - x - x^2\right)\ dx$.

SOLUTION: This represents the area under the parabolic curve $y = 12 - x - x^2$ as can be seen by the graph of the function below. The calculation of this area is as follows:

$$\left[12x - \frac{x^2}{2} - \frac{x^3}{3}\right]_{-4}^{3} = 36 - \frac{9}{2} - \frac{27}{3} - \left(-48 - \frac{16}{2} + \frac{64}{3}\right) = \frac{343}{6}.$$

This type of calculation with fractions and negatives is commonplace with definite integrals. And typically, calculators are not helpful with the arithmetic because graphing calculators can just calculate the definite integral directly. The AP testers know this so if they want to test your ability to integrate, the problem will not be calculator-active.

However if you are permitted to use your calculator to compute a definite integral, there are two ways to accomplish this as shown below: First, from the graph screen (2nd CALC 7: $\int f(x)\,dx$)), or second, directly from the home screen using MATH 9: fnInt (with the function in Y1). The newer operating system screen is shown as well.

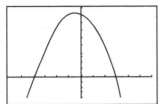

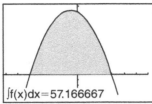

| fnInt (Y$_1$, X, −4, 3) |
| ▸ Frac |
| 343/6 |

$$\int_{-4}^{3} (12 - X - X^2)\, dX$$

57.16666667

Ans ▸ Frac

$$\frac{343}{6}$$

TEST TIP

There are two calculus applications that are available on the TI-84+ calculus – taking the derivative (using the nDeriv function) and calculating definite integrals (using the fnInt function). AP test problems requiring nDeriv are rare because of the ease of taking derivatives. But finding definite integrals using fnInt occurs all the time, mostly because of the difficulty of integrating and also the messy arithmetic. It is vital that you be able to use technology to find definite integrals on the AP exam, so be sure you know the necessary keystrokes.

EXAMPLE 21:   Find $\displaystyle\int_0^{\pi} (2\sin x + 4\cos x + 1)\, dx$.

SOLUTION: $\left[-2\cos + 4\sin x + x \right]_0^{\pi}$

$= -2\cos \pi + 4\sin \pi + \pi - (-2\cos 0 + 4\sin 0 + 0)$

$= 2 + 0 + \pi + 2 - 0 - 0 = 4 + \pi$

TEST TIP

When the lower limit of a definite integral is zero, typically the 2nd half of the integral calculation disappears. However, this does not always happen with trig functions and expressions involving e. Be very careful when doing these types of calculations, especially on the multiple-choice section of the AP exam. Typical incorrect choices are based upon this kind of careless mistake.

EXAMPLE 22: Find $\displaystyle\int_0^1\left(\sqrt{x}+\frac{1}{x+1}-e^x\right)dx$.

SOLUTION: $\left[\dfrac{2}{3}x^{3/2}+\ln(x+1)-e^x\right]_0^1$

$=\dfrac{2}{3}+\ln 2-e-(0+0-1)$

$=\dfrac{5}{3}+\ln 2-e$

This answer is negative. Definite integrals can be negative. It simply means that there is more area below the x-axis than above it.

EXAMPLE 23: Find $\displaystyle\int_{-4}^{2}|2x+4|\ dx$.

SOLUTION: This problem shows the importance of always thinking of the meaning of a definite integral: area under a curve. To do this analytically, it is necessary to write $|2x+4|$ as a piecewise function:

$$|2x+4|=\begin{cases}2x+4, x\ge -2\\ -2x-4, x< -2\end{cases}.$$

Now the definite integral has to be broken into two pieces:

$$\int_{-4}^{2}|2x+4|\ dx=\int_{-4}^{-2}(-2x-4)\ dx+\int_{-2}^{2}(2x+4)\ dx.$$

Finally each definite integral has to be evaluated:

$$\int_{-4}^{2}|2x+4|\ dx=\int_{-4}^{-2}(-2x-4)\ dx+\int_{-2}^{2}(2x+4)\ dx$$

$$\int_{-4}^{2}|2x+4|\ dx=\left[-x^2-4x\right]_{-4}^{-2}+\left[x^2+4x\right]_{-2}^{2}$$

$$=(-4+8)-(-16+16)+(4+8)-(4-8)$$
$$=4-0+12+4=20$$

Yet, this problem can be done easily by graphing the function. Students should know how to graph an absolute value curve from precalculus, but plotting points will work as well.

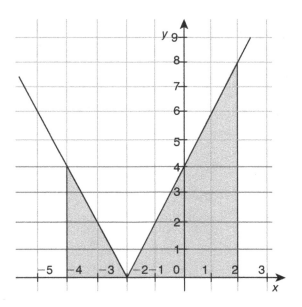

$\int\limits_{-4}^{2} |2x + 4|\ dx$ merely represents the area of two triangles:

$\int\limits_{-4}^{2} |2x + 4|\ dx = \dfrac{1}{2}(2)(4) + \dfrac{1}{2}(4)(8) = 4 + 16 = 20.$ So picturing the graph geometri-

cally can save a great deal of work.

TEST TIP

Integrals of absolute value functions show up regularly on the AP exam. When you see a definite integral, first think AREA instead of integrating and then evaluating. You may very well have to integrate the expression algebraically, but thinking AREA first will definitely save you time.

EXAMPLE 24:   If $-\dfrac{\pi}{2} < x \leq \dfrac{\pi}{2}$, for what value of x does $\displaystyle\int_0^x 4\sec t \tan t \; dt = 4$?

SOLUTION: $[4\sec t]_0^x = 4\sec x - 4\sec 0 = 4$

$4\sec x = 8$

$\sec x = 2$ so $\cos x = \dfrac{1}{2}$

$x = \pm\dfrac{\pi}{3}$

Don't forget the negative solution here. The graph of the function shows that if x is negative, the function is negative and dx is also negative, making the integral positive.

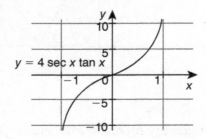

$y = 4\sec x \tan x$

EXAMPLE 25:   Let $F(x)$ be an antiderivative of $\sin^3 x$. If $F(1) = 2$, find $F(2)$.

SOLUTION: It would appear that we are being asked to find $\int \sin^3 x \; dx$, which, to our knowledge cannot be integrated. But if $F(x) = \int \sin^3 x \; dx$, it follows that $\displaystyle\int_1^2 \left(\sin^3 x\right) \; dx = F(2) - F(1)$ so $F(2) = F(1) + \displaystyle\int_1^2 \left(\sin^3 x\right) \; dx$.

We cannot integrate $\displaystyle\int_1^2 \left(\sin^3 x\right)$ by hand, but the calculator can do the computation.

```
fnInt ((sin(X))³ , X, 1 , 2
                    .8798504221
```

$F(2) \approx 2 + 0.880 = 2.880$

TEST TIP

The formula $F(b) = F(a) + \int_a^b f(x)\ dx$ where F is an antiderivative of f shows up many times in the AP exam, especially in word problems. It is an important formula that should be memorized.

DIDYOUKNOW?

The TI-84+ Calculator does not actually calculate a definite integral. Instead, it approximates the value of the definite integral to high precision, based on the type of function. For areas under lines or parabolas (quadratic functions), the calculator will find the definite integral perfectly, but other functions introduce very slight errors. Using Riemann sums or trapezoids would take too much time to give 10-place accuracy, so a much more sophisticated method called the Gauss-Kronrod quadrature formula is used to give an approximation of the integral in a few seconds. But for all functions on the AP test, your calculator will accurately give three-decimal-place accuracy. If the calculator appears to freeze while computing a definite integral, don't panic—sometimes it takes time. To stop a calculation in the middle, press the ON button.

Earlier in this chapter, we worked on problems with indefinite integrals where it was necessary to use u-substitution to perform the integration. When we are involved with these more complex expressions in order to find definite integrals, there is a technique called **changing the limits** that makes the process slightly easier.

EXAMPLE 26:   Find $\int_0^4 x\sqrt{16 - x^2}\,dx$.

SOLUTION: We start the problem like it was an indefinite integral:

$$\int_0^4 x\left(16 - x^2\right)^{1/2}\ dx \quad u = 16 - x^2,\ du = -2x\ dx$$

We need a $-2x\,dx$ in the problem so we both multiply and divide by

-2: $\dfrac{-1}{2}\displaystyle\int_0^4 -2x\left(16-x^2\right)^{1/2}\,dx$.

However, when we change everything to u's, the limits of integration are incorrect as they pertain to x.

So at this point, we have a choice: work in x or work in u. Working in u is called changing the limits. If we change the limits, we never have to switch back to x.

Working in x:

$\dfrac{-1}{2}\displaystyle\int u^{1/2}\,du = \dfrac{-1}{2}\left(\dfrac{2}{3}\right)u^{3/2}$

We now switch back to x

$\dfrac{-1}{3}\left(16-x^2\right)^{3/2}\Big|_0^4$

$\dfrac{-1}{3}(0-64)=\dfrac{64}{3}$

Changing the limits:

Change the limits: $x=0, u=16 \quad x=4, u=0$

$\dfrac{-1}{2}\displaystyle\int_{16}^0 u^{1/2}\,du = \dfrac{-1}{2}\left(\dfrac{2}{3}\right)u^{3/2}\Big|_{16}^0$

$\dfrac{-1}{3}(0-64)=\dfrac{64}{3}$

Either way, you get the same answer. Changing the limits means a bit less writing.

EXAMPLE 27: Find $\displaystyle\int_0^{\pi/2} \dfrac{\sin x}{\sqrt{\cos x}}\,dx$.

SOLUTION: $\displaystyle\int_0^{\pi/2} \sin x\,(\cos x)^{-1/2}\,dx$

$-\displaystyle\int_0^{\pi/2} -\sin x\,(\cos x)^{-1/2}\,dx \quad u=\cos x, du=-\sin x\,dx$

$\qquad\qquad\qquad\qquad\qquad\qquad x=0, u=1 \quad x=\dfrac{\pi}{2}, u=0$

$-\displaystyle\int_1^0 u^{-1/2}\,du$

We can get rid of the negative sign and switch the limits at the same time.

$\displaystyle\int_0^1 u^{-1/2}\,du$

$2u^{1/2}\Big|_0^1 = 2$

TEST TIP

Your teacher may not have taught changing the limits because it saves at most one step from simply working in x. However, there is a type of problem that has shown up on the AP exam that tests changing the limits. It usually shows up in a multiple-choice problem as illustrated in Example 28. This is reason enough to learn the technique. Calculators are useless for this problem.

EXAMPLE 28: If $F(x) = \int f(x)\, dx$, find $\int_{-2}^{4} f\left(\frac{1}{2}x\right) dx$.

(A) $\dfrac{1}{2}F(8) - \dfrac{1}{2}F(-4)$

(B) $\dfrac{1}{2}F(4) - \dfrac{1}{2}F(-2)$

(C) $2F(4) - 2F(2)$

(D) $2F(2) - 2F(-1)$

(E) $F(2) - F(-1)$

SOLUTION:

$\int_{-2}^{4} f\left(\frac{1}{2}x\right) dx$

$2\int_{-2}^{4} \frac{1}{2} f\left(\frac{1}{2}x\right) dx$

$2\int_{-1}^{2} f(u)\, du$

$2\left[F(u)\right]_{-1}^{2} = 2F(2) - 2F(-1)$ (D)

$u = \dfrac{1}{2}x,\ du = \dfrac{1}{2}dx$

$x = -2, u = -1 \quad x = 4, u = 2$

Only a student who truly understands *u*-substitution and changing the limits will get this question correct.

The Second Fundamental Theorem of Calculus

Overview: The conclusion of the **Second Fundamental Theorem of Calculus** is quite simple, but opens the door to many types of problems. It says that for a continuous function *f* on an interval containing *a*, for every *x* in the interval, $\dfrac{d}{dx}\displaystyle\int_a^x f(t)\, dt = f(x)$. In words it is saying that if you take a derivative of the accumulation function, the derivative and integral essentially cancel out and you are left with $f(x)$.

The 2nd FTC goes a step further and says that $\dfrac{d}{dx}\displaystyle\int_a^u f(t)\, dt = f(u)\cdot\dfrac{du}{dx}$. This is the chain rule at work.

EXAMPLE 29: Show that the 2nd FTC holds for $f(t) = t^2 + 3t - 9$ on $[2, x]$.

SOLUTION: The 2nd FTC says that $\dfrac{d}{dx}\displaystyle\int_2^x \left(t^2 + 3t - 9\right) dt = x^2 + 3x - 9$.

$$\int_2^x \left(t^2 + 3t - 9\right) dt = \left[\frac{t^3}{3} + \frac{3t^2}{2} - 9t\right]_2^x = \frac{x^3}{3} + \frac{3x^2}{2} - 9x - \left(\frac{8}{3} + 6 - 18\right)$$

$$\frac{d}{dx}\left[\frac{x^3}{3} + \frac{3x^2}{2} - 9x - \left(\frac{8}{3} + 6 - 18\right)\right] = x^2 + 3x - 9$$

EXAMPLE 30: Find $\dfrac{d}{dx}\displaystyle\int_1^x \sqrt[3]{t^2 + 5}\, dt$.

SOLUTION: You cannot take the integral of $\sqrt[3]{t^2 + 5}$, but you are not being asked to. Applying the 2nd FTC, you get:

$$\frac{d}{dx}\int_1^x \sqrt[3]{t^2 + 5}\, dt = \sqrt[3]{x^2 + 5}$$

EXAMPLE 31: Find $\dfrac{d}{dx}\displaystyle\int_{x^2}^{\pi/2}\cos^2 t\,dt$.

SOLUTION: First, reverse the limits: $-\dfrac{d}{dx}\displaystyle\int_{\pi/2}^{x^2}\cos^2 t\,dt$

Then, by the 2nd FTC and the chain rule you get:

$$-\frac{d}{dx}\int_{\pi/2}^{x^2}\cos^2 t\,dt=(-2x)\cos^2\left(x^2\right)$$

EXAMPLE 32: Below is the graph of $y=f(x)$ on the domain $[-4, 5]$, made up of lines. Let $g(x)=\displaystyle\int_0^x f(t)\,dt$.

a) Determine the maximum and minimum value of g on $[-4, 5]$.

b) Determine the locations of any inflection points on the graph of $g(x)$.

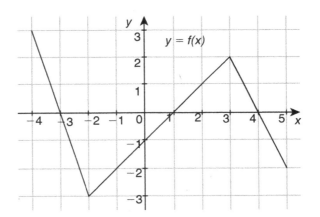

SOLUTIONS: a) This is the same problem as Example 13. However, when we examined it before, we did not have the knowledge of the 2nd FTC, so we were forced to examine $g(-4)$, $g(-3)$,..., $g(5)$ and determine which was the largest and which was the smallest. And that method did not examine the possibility that the maximum or minimum could have been based on a non-integer x-value.

The 2nd FTC makes the problem easier. We know that maximizing or minimizing an expression means taking its derivative. And we know $g'(x) = \dfrac{d}{dx}\displaystyle\int_0^x f(t)\,dt = f(x)$. Realize that we are looking at a graph of $f(x)$ above. Making a sign chart of $f(x)$, we get:

$$g'(x) = f(x): \underset{\substack{\;\;\;\;\;\; -4 \;\;\;\;\;\; -3 \qquad\qquad\quad 1 \qquad\quad 4 \qquad\quad 5}}{++++0--------0+++++0-----}$$

Interpreting the sign chart, we find that g is increasing on $[-4, -3)$ and $(1, 4)$ and decreasing on $(-3, 1)$ and $(4, 5]$.

Thus, g has relative maxima at $x = -3$ and $x = 4$.

$g(-3) = \displaystyle\int_0^{-3} f(t)\,dt = 5.5$ and $g(4) = \displaystyle\int_0^4 f(t)\,dt = 2.5$ so the maximum value of g is 5.5 at $x = -3$.

Interpreting the sign chart, we find that g has a relative minimum at $x = 1$. However, we must examine the endpoints to find the absolute minimum.

$g(1) = \displaystyle\int_0^1 f(t)\,dt = -0.5,\ g(-4) = \displaystyle\int_0^{-4} f(t)\,dt = 4,\ g(5) = \displaystyle\int_0^5 f(t)\,dt = 1.5$
so the minimum value of g is -0.5 at $x = 1$.

b) Finding inflection points on g means looking at g''. Since $g'(x)=f(x)$, it follows that $g''(x)=f'(x)$. So we make a sign chart of $g''(x)=f'(x)$.

$$g''(x) = f'(x): \underset{\substack{\;\;\;\;\;\; -4 \qquad\qquad\; -2 \qquad\qquad\quad 3 \qquad\qquad\quad 5}}{--------0+++++++++++0---------}$$

Interpreting the sign chart, we find that g is concave down on $[-4, -2)$ and $(3, 5]$ and concave up on $(-2, 3)$. So g has inflection points at $x = -2$ and $x = 3$.

TEST TIP

For many students, these types of AP exam problems are the most difficult ones they encounter. There is a lot of information given and it is easy to get confused. When you see a problem like the following one, break it down into little steps and basic definitions. Realize also that a problem with subparts tests many concepts while a multiple-choice problem is more likely to test only one concept.

EXAMPLE 33: Let $f(x)$ be defined by the graph below whose domain is $[-4, 4]$. Let $F(x) = \int_{-3}^{x} f(t)\, dt$.

a) Find the equation of the tangent line to F at $x = 3$.

b) Use the results from part a) to approximate the value of F at $x = 3.1$. Does this value over-approximate or under-approximate $F(3.1)$. Justify your answer.

c) Find the value of x where F has its minimum value. Justify your answer.

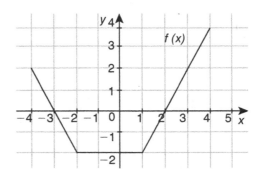

SOLUTIONS: a) When you are asked to find the equation of a tangent line, you need the formula: $y - y_1 = m(x - x_1)$. So you need 2 pieces of information: $F(3)$ and $F'(3)$.

By definition, $F(3) = \int_{-3}^{3} f(t)\, dt$. This is the area "under" the graph of f starting at -3 and ending at 3. Your dt is positive and the function is negative, then positive, so $F(3) = -7$.

$F'(3)$ is easier to find. By the 2nd FTC, $F'(x) = \dfrac{d}{dx} \int_{-3}^{x} f(t)\, dt = f(x)$, so $F'(3) = f(3) = 2$.

Put it together. Tangent line: $y + 7 = 2(x - 3)$.

b) $y + 7 = 2(3.1 - 3)$ so $y \approx -6.8$. Since F is increasing at $x = 3$ (as $F'(3) > 0$) and concave up at $x = 3$ (as $F''(3) = f'(3) > 0$), the value of -6.8 under-approximates $F(3.1)$.

c) Make a sign chart of F' which (by the 2nd FTC) is the same as f.

$$F' = f : \underset{\substack{\\ -4 \quad -3 \qquad\qquad\qquad\qquad 2 \qquad\qquad 4}}{+\!+\!+\,0\,-\!-\!-\!-\!-\!-\!-\!-\!-\!-\!-\,0\,+\!+\!+\!+\!+\!+}$$

Since F is increasing on $[-4, -3)$, it could have its minimum value at $x = -4$. And since F' switches from negative to positive at $x = 2$, F has a relative minimum at $x = 2$.

$$F(-4) = \int_{-3}^{-4} f(t)\, dt = -1 \quad \text{and} \quad F(2) = \int_{-3}^{2} f(t)\, dt = -8.$$

So the minimum value of F on $[-4, 4]$ is -8.

TEST TIP

While doing analysis of the first derivative will eliminate some possible locations of maximum or minimum values, it is permissible on the AP exam to simply evaluate all critical values and endpoints to find the largest and smallest. When you do so, put the results in a chart to make it easier for the exam readers to follow.

EXAMPLE 34:   The functions f and g are differentiable for all real numbers. The table below gives values of the functions and their first derivatives at selected values of x. Let h be the function given by $h(x) = \displaystyle\int_{-1}^{g(2x)} f(t)\, dt$. Find the equation of the tangent line to h at $x = 1$.

x	$f(x)$	$f'(x)$	$g(x)$	$g'(x)$
−1	3	0	−2	1
1	6	−2	−2	−1
2	3	−4	−1	−5

SOLUTION: $h(1) = \displaystyle\int_{-1}^{g(2)} f(t)\, dt = \int_{-1}^{-1} f(t)\, dt = 0$

By the 2nd FTC and the chain rule, $h'(x) = f\big(g(2x)\big) \cdot g'(2x)(2)$

$h'(1) = f\big(g(2)\big) \cdot g'(2)(2) = f(-1) \cdot (-5)(2) = 3(-5)(2) = -30$

Tangent line: $y - 0 = -30(x - 1)$ or $y = 30 - 30x$

Time for a quiz
• Review strategies in Chapter 2
• Take Quiz 5 at the REA Study Center
    (www.rea.com/studycenter)

Applications of Integration

Area Between Curves

Overview: We have shown that the area under a curve on [a, b] is given by $\int_a^b f(x)\,dx$. If we have two continuous functions $f(x)$ and $g(x)$ with $f(x) \geq g(x)$, then using vertical rectangles, the area between f and g and the vertical lines $x = a$ and $x = b$ is given by $\int_a^b [f(x) - g(x)]\,dx$. It makes no difference whether the functions are above or below the x-axis.

The formula above can be amended to work with horizontal rectangles as well (when it is more convenient to work with functions in terms of y instead of x). The best way to remember the **area between curve** formulas is to follow these "recipes":

Vertical rectangles:

$$\text{Area} = \int_{x_1}^{x_2} [\text{top curve} - \text{bottom curve}]\, dx$$

$$= \int_{x_1}^{x_2} [t(x) - b(x)]\, dx$$

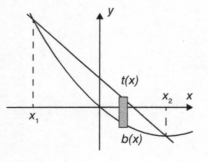

Horizontal rectangles:

$$\text{Area} = \int_{y_1}^{y_2} [\text{right curve} - \text{left curve}]\, dy$$

$$= \int_{y_1}^{y_2} [r(y) - l(y)]\, dy$$

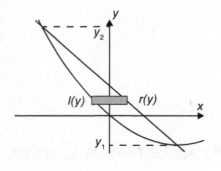

EXAMPLE 1: Find the area between the curves $y = 8 + x - x^2$ and $y = 2x - 4$.

SOLUTION: This is not a calculator-active question, so some knowledge of functions and algebra is required. Without spending the time to actually graph the functions, you should know that $y = 8 + x - x^2$ is a parabola opening down and $y = 2x - 4$ is a line with positive slope. So the graph (without axes or scale) must appear like this:

With it being clear which function is on top and which is on the bottom, we need only find the limits of integration. We do that by setting the functions equal to each other.

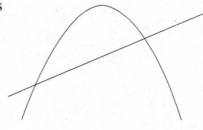

$$8 + x - x^2 = 2x - 4$$
$$0 = x^2 + x - 12$$
$$(x + 4)(x - 3) = 0$$
$$x = -4, x = 3$$

Finally, we can set up the integral and use the Fundamental Theorem to calculate the area.

$$A = \int_{-4}^{3} \left[8 + x - x^2 - (2x - 4) \right] dx$$

$$A = \int_{-4}^{3} \left(12 - x - x^2 \right) dx$$

$$A = \left[12x - \frac{x^2}{2} - \frac{x^3}{3} \right]_{-4}^{3} = 36 - \frac{9}{2} - 9 - \left(-48 - 8 + \frac{64}{3} \right)$$

$$A = 83 - \frac{9}{2} - \frac{64}{3} = \frac{343}{6}$$

TEST TIP

Problems similar to the one above take a lot of time, usually involve fractions, and there are many opportunities to make careless errors. If calculators are not permitted, usually students will be asked to set up, but not actually calculate, the integral.

But typically, area problems (combined with volume, covered later in this chapter) show up in a free-response question on the calculator-active section of the AP exam. Students are often asked to set up an integral representing the area between functions and then possibly calculate it. Be sure you know how to use your calculator to 1) graph functions, 2) find intersection of functions (2nd CALC 5:intersect), and 3) find definite integrals, even if you can integrate using the Fundamental Theorem. Input your functions in Y1 and Y2 and use the command MATH 9:fnINT(Y1-Y2, X, lower limit, upper limit).

EXAMPLE 2: (Calculator Active) Find the area bounded by the curves $y = 6\sin x$ and $y = 4\ln(x - 2) - 3$.

SOLUTION: This is a problem that will be done completely using the calculator. Realizing that for the second function, the domain is $x > 2$ makes finding an appropriate graphing window a bit easier. It should be obvious that there are two areas that need to be found. The calculator is used to find the x-values of the three intersection points (3.414, 6.872 and 8.569) and the definite integrals are written and computed.

$$A = \int_{3.414}^{6.872} \left[4\ln(x-2) - 3 - 6\sin x \right] dx + \int_{6.872}^{8.569} \left[6\sin x - (4\ln(x-2) - 3) \right] dx$$

$$A \approx 15.462 + 2.201 = 17.663$$

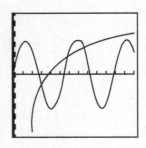

Plot1 Plot2 Plot3	WINDOW
\Y$_1$ = 6sin(X)	Xmin = 0
\Y$_2$ = 4ln(X−2)−3	Xmax = 15
\Y$_3$ =	Xsc1 = 1
\Y$_4$ =	Ymin = 10
\Y$_5$ =	Ymax = 10
\Y$_6$ =	Ysc1 = 1
\Y$_7$ =	Xres = 1

fnInt (Y$_2$ − Y$_1$, X, 3.414,6.872) + fnInt (Y$_2$ − Y$_2$, X, 6.872,8.569)

17.66311106

EXAMPLE 3: The figure below shows the graphs of $x = 1 - y^2$ and $x = 1 - y$. Find the area between the curves.

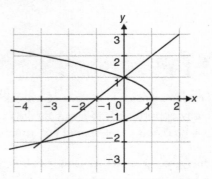

SOLUTION: This problem could be done with the vertical rectangle or horizontal rectangle technique:

Vertical Rectangles

$y^2 = 1 - x \Rightarrow y = \pm\sqrt{1-x}$

$x = y - 1 \Rightarrow y = x + 1$

There are 2 sections so 2 integrals are needed.

$$A = \int_{-3}^{0} \left[(x+1) - \left(-(1-x)^{1/2}\right)\right]dx + \int_{0}^{1}\left[(1-x)^{1/2} - \left(-(1-x)^{1/2}\right)\right]dx$$

This is a tough calculation and will take time.

Using Horizontal Rectangles is easier.

Horizontal Rectangles

$$A = \int_{-2}^{1}\left[1 - y^2 - (y-1)\right]dy$$

$$A = \left[2y - \frac{y^2}{2} - \frac{y^3}{3}\right]_{-2}^{1}$$

$$A = 2 - \frac{1}{2} - \frac{1}{3} - \left(-4 - 2 + \frac{8}{3}\right)$$

$$A = 8 - \frac{1}{2} - 3 = \frac{9}{2}$$

It should be clear that using horizontal rectangles make the integration calculation much easier.

TEST TIP

If a calculator is allowed for the previous problem on the AP exam and the horizontal rectangle technique is used, even though the integration is in terms of *y*, you may use the *x*-variable on the calculator.

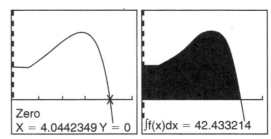

> fnInt (1 − X² − (X−1)
> , X, ⁻2, 1) ▸ Frac
>   9/2

EXAMPLE 4: (Calculator Active) Given $f(x) = 8 + 3x^2 - e^x$, find the value of k such that the line $x = k$ divides the area between $f(x)$, the *x*-axis, and the *y*-axis into two equal areas.

SOLUTION: First graph $f(x)$ and find where it crosses the *x*-axis to get a feel for the problem and compute the area.

Zero
X = 4.0442349 Y = 0   ∫f(x)dx = 42.433214

Set up an integral in terms of k that describes the problem situation:

$$\int_0^k \left(8 + 3x^2 - e^x\right)\,dx = \frac{42.433}{2}$$

Note: an alternative equation is $\displaystyle\int_0^k \left(8 + 3x^2 - e^x\right)dx = \int_k^{4.044} \left(8 + 3x^2 - e^x\right)dx$

$$\left[8x + x^3 - e^x\right]_0^k = 21.217$$

$$8k + k^3 - e^k - (-1) = 21.217$$

$$8k + k^3 - e^k - 20.217 = 0$$

Now graph that equation and find the solution on the interval [0, 4.044].

The solution is $k \approx 2.273$. You may want to check out that the solution is correct by integrating from 0 to k and then from k to 4.044 as shown on the screen below. Neglecting rounding errors, the two areas are equal.

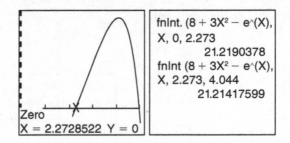

fnInt. $(8 + 3X^2 - e\char`^(X))$,
X, 0, 2.273
                21.2190378
fnInt $(8 + 3X^2 - e\char`^(X))$,
X, 2.273, 4.044
                21.21417599

Zero
X = 2.2728522 Y = 0

Average Value of a Function

Overview: We have a Mean-Value Theorem for Derivatives. There is also one for integrals. If we have a continuous function f on an interval $[a, b]$, the **Mean-Value Theorem for Integrals** states that there exists a c on $[a, b]$ such that $\int_{a}^{b} f(x) \ dx = f(c)(b-a)$. In words, this says that the area under the curve $f(x)$ between a and b can be expressed as the area of a rectangle where the base is $b - a$ and the height is $f(c)$. (See diagram below.) This $f(c)$ is called the **average value of the function.**

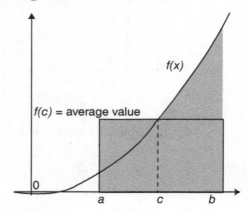

In mathematics, an average is a measure of the middle or typical value of a set of data. When you find your average in your calculus class, you come up with a number that describes your typical score. To do this, you add up all your exam scores and divide by the number of exam scores (assuming that they are all based on the same number of points). If we are asked to find the average temperature on a typical day, we might add the temperatures every hour and divide by 24. However, if there are big fluctuations between measurements, your average temperature calculation might be off.

So when you have a continuous function $f(x)$ and wish to find the **average value of the function** f_{avg} on an interval $[a, b]$, we sum the area under the curve and divide by $b - a$. This follows from the Mean-Value Theorem for Integrals:

$$\int_a^b f(x)\ dx = f(c)(b-a) \quad \text{so} \quad f(c) = f_{avg} = \frac{\int_a^b f(x)}{b-a}dx.$$

The units for the average value will be the same units as the function f.

Students typically confuse the average value of a function over $[a, b]$ with the average rate of change of a function over $[a, b]$. Although these are different calculations, they are based on the same formula.

Average value of a function $f(x)$ over $[a, b]$: $f_{avg} = \dfrac{\int_a^b f(x)}{b-a}dx$

Average rate of change of a function $f(x)$ over $[a, b]$: $\dfrac{f(b)-f(a)}{b-a}$. But realize that this is the same thing as the average value of the rate of change of f: $\dfrac{\int_a^b f'(x)dx}{b-a}$.

EXAMPLE 5: If $f(x) = 4x^3 - 6x^2 - 2x - 1$, find
a) the average value of f on $[-1, 1]$
b) the average rate of change of f on $[-1, 1]$

SOLUTIONS:

a) $f_{avg} = \dfrac{\int_{-1}^{1}\left(4x^3 - 6x^2 - 2x - 1\right)\ dx}{1+1} = \dfrac{\left(x^4 - 2x^3 - x^2 - x\right)\Big|_{-1}^{1}}{2} = \dfrac{-3-3}{2} = -3$

b) $\dfrac{f(1)-f(-1)}{1+1} = \dfrac{4-6-2-1-(-4-6+2-1)}{2} = \dfrac{-5+9}{2} = 2$

Realize that you could have used the average value function (taking the average value of the rate of change of f)

$$\dfrac{\int_{-1}^{1}\left(12x^2 - 12x - 2\right)dx}{1+1} = \dfrac{\left(4x^3 - 6x^2 - 2x - 1\right)\Big|_{-1}^{1}}{2} = \dfrac{4-6-2-1-(-4-6+2-1)}{2} = \dfrac{-5+9}{2} = 2$$

EXAMPLE 6: Find the average value of $f(x) = \sqrt{\sin x} \cos x \, dx$ on $\left[\dfrac{\pi}{2}, \pi\right]$.

SOLUTION:

$$f_{avg} = \frac{\displaystyle\int_{\pi/2}^{\pi} (\sin x)^{1/2} (\cos x)}{\pi - \dfrac{\pi}{2}} = \frac{\left[\dfrac{2(\sin x)^{3/2}}{3}\right]_{\pi/2}^{\pi}}{\pi - \dfrac{\pi}{2}} \quad u = \sin x, \, du = \cos x \, dx$$

$$\frac{\left[\dfrac{\dfrac{2}{3}(0-1)}{\pi - \dfrac{\pi}{2}}\right]\left(\dfrac{6}{6}\right) = \frac{-4}{6\pi - 3\pi} = \frac{-4}{3\pi}$$

EXAMPLE 7: Let $f(x) = 3x^2 + 2x + 1$. For what value(s) of a does the average value of $f(x)$ on the interval $[0, a]$ equal the average rate of change of $f(x)$ on the interval $[0, a]$?

SOLUTION:
$$\frac{\displaystyle\int_{0}^{a} (3x^2 + 2x + 1) \, dx}{a - 0} = \frac{(3a^2 + 2a + 1) - 1}{a}$$

$$\frac{\left[x^3 + x^2 + x\right]_{0}^{a}}{a} = 3a + 2$$

$$a^2 + a + 1 = 3a + 2$$

$$a^2 - 2a - 1 = 0$$

$$a = \frac{2 \pm \sqrt{4+4}}{2} = 1 \pm \sqrt{2}$$

EXAMPLE 8: Tim takes 60 seconds to park his car. His velocity $v(t)$ at time t, $0 \le t \le 60$, given in ft/sec is given by the function whose graph is given below. What is Tim's average speed when he parks his car?

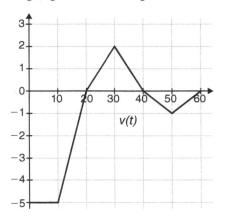

SOLUTION:

$$\text{Average Speed} = \frac{\int_0^{60} |v(t)|\, dt}{60 - 0} = \frac{-\int_0^{10} v(t)\, dt - \int_{10}^{20} v(t)\, dt + \int_{20}^{40} v(t)\, dt - \int_{40}^{60} v(t)\, dt}{8 - 0}$$

$$= \frac{50 + 25 + 20 + 10}{60} = \frac{105}{60} = \frac{7}{4} = 1\frac{3}{4}\ \frac{\text{ft}}{\text{sec}}$$

EXAMPLE 9: (Calculator Active) The temperature of ocean water at Ocean City, N.J., is modeled by the equation $T(t) = 56 + 19\cos\left[\dfrac{2\pi(t - 220)}{365}\right]$

where t is the day of the year and T is measured in degrees Fahrenheit. The summer season is generally considered to be between Memorial Day (day 150) and Labor Day (day 245).

a) What is the average water temperature during the summer season at Ocean City?

b) It is estimated that Ocean City beach vendors will sell 500 ice cream treats for every degree the water temperature is above 62°F. These treats give a profit of 79 cents per treat. How much profit will the vendors make during the summer season?

SOLUTIONS: a) $T_{avg} = \dfrac{\int_{150}^{245}\left(56 + 19\cos\left[\dfrac{2\pi(t - 220)}{365}\right]\right) dt}{245 - 150} \approx 71.697°\ \text{F.}$

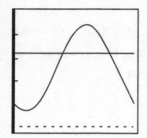 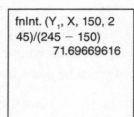 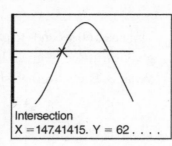 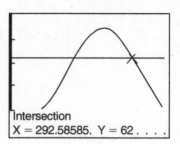

b) We first find the days when the water temperature is above 62 degrees.

The area between the curve and the line represents the total accumulation of degrees for all days during the year that the temperature of the water is above 62 degrees.

As shown by the graphs above, the water temperature is above 62° between days 147 and 293. But since the summer season ends on Labor Day (day 245), we only integrate to 245.

$$\text{Ice cream treats} \approx 500 \int_{147}^{245} \left(56 + 19\cos\left[\frac{2\pi(t-220)}{365} \right] - 62 \right) dt$$

$$= 500(922.19) = 461096$$

$$\text{Profit} \approx 0.79(461096) \approx \$364,266$$

DIDYOUKNOW?

The Great Pyramid in Giza, Egypt, was built with such great precision that modern technology cannot duplicate it. There are many fascinating facts about the pyramid, but one relates to the average value function: the average height of land above the earth is 5,449 inches. This is also the height of the pyramid.

Accumulated Change Problems

Overview: We now combine two topics, one from differential calculus, the other from integral calculus to solve word problems that deal with **accumulated change**. A quantity that is given as a rate of change is interpreted as a derivative $R'(t)$ with respect to time of some function. If we want to find the total rate of change over some interval $[t_1, t_2]$, we use the Fundamental Theorem of Calculus to say: $\int_{t_1}^{t_2} R'(t)\, dt = R(t_2) - R(t_1)$. This is equivalent to saying: $R(t_2) = R(t_1) + \int_{t_1}^{t_2} R'(t)\, dt$.

EXAMPLE 10: (Calculator Active) A roller coaster breaks down temporarily and a line forms to ride it once it re-opens. At the time that the ride re-opens, there are 400 people waiting in line. Over the next 30 minutes when the roller coaster is running, the rate of change of people in line was given by $w(t) = 0.02(t^3 - 22.5t^2 - 15t + 500)$ people/minute where t is measured in minutes.

a) What percentage of time over the 30-minute period was the line getting shorter?

b) During the period of time that the line was getting shorter, how many fewer people were in line?

c) During the 30-minute period, what is the maximum number of people in line?

SOLUTIONS: a) It is important to understand that although the given equation is for $w(t)$, it is clear that $w(t)$ is a rate. So we graph it. The graph of $w(t)$ is below. We need to find the two times between which $w(t)$ is negative between which is the interval of time when the line is getting shorter. This interval is [4.923, 22.159]. So the line is getting shorter $\dfrac{22.159 - 4.923}{30} \approx 57.5\%$ of the time.

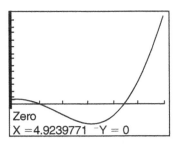
Zero
X =4.9239771 Y = 0

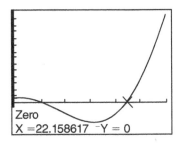
Zero
X =22.158617 Y = 0

b) Since the x-axis is measured in minutes and $w(t)$ is measured in people/minute, the definite integral (area) will be measured in people. $\displaystyle\int_{4.923}^{22.159} w(t)\,dt \approx -309$, so there are 309 fewer people in line.

c) If we want to maximize the number of people in line, we need to look at the rate of change of people in line which is $w(t)$ and that is given. $w(t)$ switches from positive to negative at $t = 4.923$. However, the maximum value of people in line could also occur at $t = 30$. The number of people at any time is given by $400 + \int_0^k w(t)\ dt$.

time	People in line
4.923	$400 + \int_0^{4.923} w(t)dt \approx 430$
30	$400 + \int_0^{30} w(t)dt \approx 565$

Thus, the maximum number of people in line occurs at the end of the 30-minute period and is 565 people.

Note: It is permissible on the AP exam to evaluate $400 + \int_0^k w(t)dt$ at all critical points and endpoints and choose the maximum. That would involve finding the number of people in line at $k = 0$, which is 400, and at $k = 22.159$, which is approximately 121, in addition to finding the number of people in line at $k = 4.923$ and $k = 30$ as shown in the table above.

TEST TIP

On the AP exam calculator-active section, functions are frequently given on a specific domain necessitating that graphs be drawn. It can be a challenge to come up with an appropriate graph showing all of the function's behavior, wasting time that you do not have. The best way to accomplish this quickly is to input the function, set the given domain using XMIN and XMAX and then press ZOOM 0:ZoomFit. After a brief pause, the calculator will show a complete graph, configuring YMIN and YMAX to fill the screen.

EXAMPLE 11: At a summer camp, one night a week, the campers have a cook-out which lasts 45 minutes. $C(t)$ (shown below) represents the rate at which hamburgers are pulled off the grill. $E(t)$ (shown below) represents the rate at which campers are served the hamburgers, all of which will be eaten. Both graphs are measured in hamburgers/min. Hamburgers that are grilled and not immediately served are placed in a warming tray. At the start of the cookout ($t = 0$), there are 20 hamburgers already in the warming tray.

a) How many hamburgers are eaten during the cookout? (to the nearest hamburger).

b) For $0 \leq t \leq 45$, find the time intervals during which the number of hamburgers in the warming tray is decreasing. Give a reason for your answer.

c) For $0 \leq t \leq 45$, at what time t is the number of hamburgers in the warming tray the least? To the nearest hamburger, compute the number of hamburgers in the tray at this time. Justify your answer.

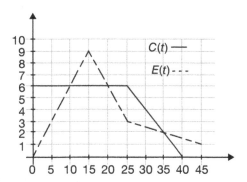

SOLUTIONS:

a) $\int\limits_{0}^{45} E(t)\,dt = \int\limits_{0}^{15} E(t)\,dt + \int\limits_{15}^{25} E(t)\,dt + \int\limits_{25}^{45} E(t)\,dt$

$\int\limits_{0}^{45} E(t)\,dt = \frac{1}{2}(15)(9) + \frac{1}{2}(10)(9+3) + \frac{1}{2}(20)(3+1) = 167.5$

168 hamburgers are eaten.

b) The number of hamburgers in the warming tray is decreasing on the intervals $10 < t < 20$ and $35 < t < 45$ because $E(t) > C(t)$ on those intervals. (Campers are eating faster than the hamburgers are being cooked.)

c) Since $C(t) - E(t)$ changes from negative to positive at $t = 20$, the candidates for the absolute minimum are at $t = 0, 20, 45$.

t(min)	hamburers in tray
0	20
20	$\int_0^{20} \left[C(t) - E(t) \right] dt = 20 + 120 - 67.5 - 30 - 7.5 = 35$
45	$\int_0^{45} \left[C(t) - E(t) \right] dt = 20 + 150 + 45 - 167.5 \approx 48$

The minimum number of hamburgers in the tray occurs at $t = 0$ minutes (start of cook-out) when there are about 20 hamburgers in the tray.

TEST TIP

From 2000 to 2012, the College Board has released 24 AP Calculus AB/BC exams. Seventeen of these exams (about 71%) have included a problem similar to the last two on the topic of accumulated change. It is highly likely that you will see one on your exam.

Straight-Line Motion Revisited

Overview: In Chapter 6, we looked at straight-line motion from the derivative point-of-view. The relationship between position $x(t)$, velocity $v(t)$, and acceleration $a(t)$ is:

$$v(t) = x'(t) \text{ and } a(t) = v'(t) = x''(t)$$

We reverse the process and express these relationships through integration. But complicating matters is the constant of integration:

$$v(t) = \int a(t)\, dt + C \quad \text{and} \quad x(t) = \int v(t)\, dt + C.$$

In terms of definite integrals, we also know that

$$v(t_2) - v(t_1) = \int_{t_1}^{t_2} a(t)\,dt \quad \text{or} \quad v(t_2) = v(t_1) + \int_{t_1}^{t_2} a(t)\,dt$$

$$x(t_2) - x(t_1) = \int_{t_1}^{t_2} v(t)\,dt \quad \text{or} \quad x(t_2) = x(t_1) + \int_{t_1}^{t_2} v(t)\,dt$$

Two other concepts that come into play are displacement and distance over some time interval $[t_1, t_2]$.

Displacement: the difference in position over $[t_1, t_2]$.

Displacement $= x(t_2) - x(t_1) = \int_{t_1}^{t_2} v(t)\,dt$. Displacement can be positive, negative, or zero.

Distance: how far the particle traveled over $[t_1, t_2]$.

Distance $= \int_{t_1}^{t_2} |v(t)|\,dt$. Distance is never a negative number.

EXAMPLE 12: A particle moves along the x-axis so that at any time $t \geq 0$, its velocity is given by $v(t) = 8 - 2t - \sin\left(\dfrac{\pi t}{2}\right)$. If the particle is at position $x = -2$ at time $t = 0$, where is the particle at $t = 1$?

SOLUTION: $x(t) = \int \left(8 - 2t - \sin\left(\dfrac{\pi t}{2}\right) \right) dt = 8t - t^2 + \dfrac{2}{\pi}\cos\left(\dfrac{\pi t}{2}\right) + C$

$$x(0) = 0 - 0 + \dfrac{2}{\pi} + C = -2 \Rightarrow C = -2 - \dfrac{2}{\pi}$$

$$x(t) = 8t - t^2 + \dfrac{2}{\pi}\cos\left(\dfrac{\pi t}{2}\right) - 2 - \dfrac{2}{\pi}$$

$$x(1) = 8 - 1 + 0 - 2 - \dfrac{2}{\pi} = 5 - \dfrac{2}{\pi}$$

EXAMPLE 13: (Calculator Active) A particle moves along the *x*-axis so that at any time $t > 0$, its acceleration is given by $a(t) = \ln(8^t + t) - 1$. If the velocity of the particle is -1 at time $t = 0$, find the speed of the particle at $t = 1$.

SOLUTION:

$$v(1) = v(0) + \int_0^1 a(t)\,dt = -1 + 0.172 = -0.828 \Rightarrow \text{Speed} = \left| -0.828 \right| = 0.828$$

```
fnInt (1n(8^X + X)−
1, X, 0, 1)
            .1720860777
```

EXAMPLE 14: A storage facility has a straight conveyor belt on which very heavy objects are placed. The belt can move objects forward or backwards to different locations called bins. A piece of machinery is being repositioned in the facility and is placed on the belt at location Bin 15. It takes 12 minutes to get to its final position. Its velocity in bins/minute is given by the function whose graph is shown below.

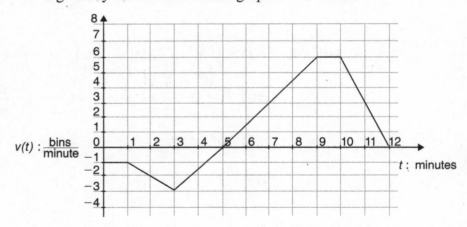

a) For $0 \le t \le 12$, at what time is the piece of machinery farthest from its starting position?

b) Where is the machinery at this time?

c) For $0 \le t \le 12$, how many bins does the piece of machinery pass?

SOLUTIONS: a) Since the velocity is zero at $t = 5$, we find the displacement of the machinery at $t = 5$ and $t = 12$.

t	Displacement: $\int_0^t v(x)dx$
5	-8
12	$-8 + 24 = 16$

The machinery is 8 bins left of its starting position at $t = 5$ and 16 bins right of its starting position at $t = 12$, so the farthest the machinery is from its starting position occurs at $t = 12$.

b) The position at $t = 12$ is $15 + \int_0^{12} v(t)dt = 15 + 16 = 31$.

c) Distance $= \int_0^{12} |v(t)|dt = -\int_0^5 v(t)dt + \int_5^{12} v(t)dt$

$-(-8) + 24 = 8 + 24 = 32$.

The machinery traveled past 32 bins.

Note that the starting position doesn't affect the answer.

EXAMPLE 15: A particle moves along the x-axis so that its velocity $v(t) = e^{-t} - \dfrac{1}{e}$, as shown in the figure below. Find the total distance that the particle traveled from $t = 0$ to $t = 2$.

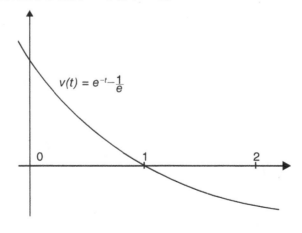

$v(t) = e^{-t} - \dfrac{1}{e}$

SOLUTION:

$$\text{Distance} = \int_0^2 \left| e^{-t} - \frac{1}{e} \right| dt = \int_0^1 \left(e^{-t} - \frac{1}{e} \right) dt - \int_1^2 \left(e^{-t} - \frac{1}{e} \right) dt$$

$$= \left[-e^{-t} - \frac{1}{e}t \right]_0^1 - \left[-e^{-t} - \frac{1}{e}t \right]_1^2$$

$$= \left(-\frac{1}{e} - \frac{1}{e} \right) - (-1 - 0) - \left[\left(-\frac{1}{e^2} - \frac{2}{e} \right) - \left(-\frac{1}{e} - \frac{1}{e} \right) \right]$$

$$= -\frac{2}{e} + 1 + \frac{1}{e^2} + \frac{2}{e} - \frac{2}{e}$$

$$= 1 - \frac{2}{e} + \frac{1}{e^2}$$

TEST TIP

A problem like Example 15 would most likely occur in the multiple-choice section and not be calculator-active. All five answer choices would look similar and with all the negative signs, the chance for errors is great. It is recommended that you calculate the two areas independently. The first (from 0 to 1) should be positive and the second (from 1 to 2) should be negative. Then subtract the second area from the first. Since no one will see your work in a multiple-choice problem, do it in an organized manner. Do it the least confusing way for yourself!

Volume

Overview: Volume problems come in two types: i) disk/washer problems rotating an area about a line and ii) cross-section problems.

The **disk/washer method** is used when rotating an area about a line of rotation. We find the outside radius R, the distance from the line of rotation to the outside curve, and, if it exists, the inside radius r, the distance from the line of rotation to the inside curve. If there is just an R, it is a disk problem. If both R and r exist, it is a washer problem.

The formula when rotating these curves about a line on an interval is given by:

$$\text{Disks: } V = \pi \int\limits_{x=a}^{x=b} \left[R(x)\right]^2 dx \quad \text{or} \quad V = \pi \int\limits_{y=c}^{y=d} \left[R(y)\right]^2 dy$$

$$\text{Washers: } V = \pi \int\limits_{x=a}^{x=b} \left(\left[R(x)\right]^2 - \left[r(x)\right]^2 \right) dx \quad \text{or} \quad V = \pi \int\limits_{y=c}^{y=d} \left(\left[R(y)\right]^2 - \left[r(y)\right]^2 \right) dy$$

When rotating around a horizontal line, the radii will be in terms of x and the limits of integration are x-values.

When rotating around a vertical line, the radii will be in terms of y and the limits of integration are y-values.

Here are the steps to follow when tackling a volume problem:

a) Draw a sketch of the function (s) and the line of rotation.

b) Determine whether it is a disk problem or a washer problem. If a disk, draw R. If a washer, draw R and r.

c) Find expressions for R and r in terms of x and y.

d) If you are rotating around a horizontal line, the integral will use dx, so change all variables to x.
If you are rotating around a vertical line, the integral will use dy, so change all variables to y.

e) Write integrals using the formulas above.

Cross-section problems create a solid with a region being the base of the solid with cross sections of the solid perpendicular to an axis as geometric shapes. These cross sections are typically squares, equilateral triangles, right triangles, or semi-circles. It is suggested that you draw the two-dimensional figure, establish its area in terms of x or y, and integrate that expression on the given interval.

EXAMPLE 16: If the region bounded by the y-axis, the line $y = 1$, and the curve $y = x^3$ is revolved about the x-axis, find the volume of the solid generated.

SOLUTION: Students should be able to sketch the function. While not necessary, sketching allows students to easily determine whether it is a disk or a washer problem and write expressions for R and r.

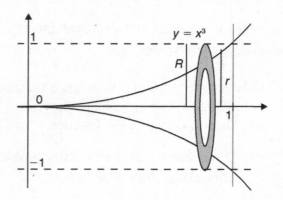

$$R = 1, r = y = x^3$$

$$x^3 = 1 \Rightarrow x = 1$$

$$V = \pi \int_0^1 \left(1 - x^6\right) dx$$

$$V = \pi \int_0^1 \left[1^2 - (x^3)^2\right] dx = \pi \int_0^1 (1 - x^6) \, dx$$

TEST TIP

You are permitted to bring several pencils into the AP exam. I recommend several colored pencils for problems like these, whether they are on the free-response or multiple-choice section. Graphing the functions and line of rotations in different colors allows you to easily determine R and r. Both R and r are always drawn from the center of rotation to the color. If you don't have colored pencils, draw the graphs with solid and dashed lines.

TEST TIP

The challenge for students doing volume problems in the AP exam is to generate expressions for R and r. A general technique for that process is to choose a point on the curve and label it as (x, y). Then write expressions for R and r in terms of x and y. In the following figure, a point has coordinates (x, y). Notice the distance from P to the x-axis is y and the distance from P to the y-axis is x. The distance from P to the line $x = 5$ is $5 - x$ and the distance from P to the line $y = 2$ is $2 - y$. Learn this technique and volume problems become easy.

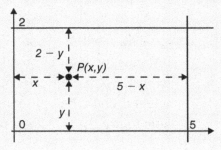

EXAMPLE 17: Let A be the region bounded by the y-axis, the function $y = \sin x$ and the line $y = 1$. Write, but do not evaluate an expression that calculates the volume of the solid generated when A is rotated about

a) the line $y = 1$.

b) the y-axis.

SOLUTIONS: a) Again, since this is not a calculator-active problem, you need to be able to draw the boundaries, identify that it is a disk problem, write an expression for R, and then write an integral.

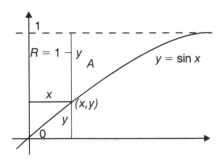

$$R = 1 - y = 1 - \sin x$$

$$V = \pi \int_0^{\pi/2} (1 - \sin x)^2 \, dx$$

b) $R = x = \sin^{-1} y$

$$V = \pi \int_0^1 \left(\sin^{-1} y \right)^2 dy$$

There will be problems where computing the integral is too difficult (or even impossible) so you only need write an integral expression. If both parts of Example 15 were calculator-active, the integral could be approximated, but then students would also be able to use the calculator to graph the problem.

EXAMPLE 18: (Calculator Active). Find the volume of the solid created when the area between $y = 4 \cos(3x) - x$ and $y = 3x^2 + e^x$ is rotated about the x-axis.

SOLUTION:

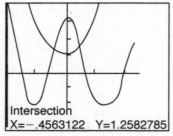
Intersection
X=−.4563122 Y=1.2582785

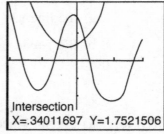
Intersection
X=.34011697 Y=1.7521506

fnInt $(Y_1{}^2 - Y_2{}^2, X,$
$-.456, .340)$
                7.19369791

$$R = 4 \cos(3x) - x, \; r = 3x^2 + e^x$$

$$V = \pi \int_{-0.456}^{0.340} \left(\left[4\cos(3x) - x \right]^2 - \left[3x^2 + e^x \right]^2 \right) dx$$

$$V = 7.194\pi$$

EXAMPLE 19: Let R be the region bounded by the graphs $y = x$ and $y = \sqrt[4]{x}$. The region R is the base of a solid. For this solid, the cross sections perpendicular to the x-axis are squares. Find the volume of this solid.

SOLUTION: Again, not being calculator active, students should be able to sketch the curves and realize they intersect at $x = 0$ and $x = 1$.

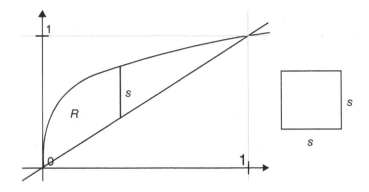

$$\text{Area} = s^2 = \left(x^{1/4} - x \right)^2$$

$$\text{Volume} = \int_0^1 \left(x^{1/4} - x \right)^2 dx$$

$$= \int_0^1 \left(x^{1/2} - 2x^{5/4} + x^2 \right) dx$$

$$= \left[\frac{2}{3} x^{3/2} - 2 \left(\frac{4}{9} \right) x^{9/4} + \frac{x^3}{3} \right]_0^1 = \frac{2}{3} - \frac{8}{9} + \frac{1}{3} = \frac{1}{9}$$

Notice that there is no π in the integral because the cross sections are squares.

TEST TIP

If Example 19 were part of a free-response question on the AP exam, a sketch of the curves most likely will be provided because if the students cannot graph it, they have no chance of receiving any of the points for the problem. The problem is testing the ability to find volume and not specifically graphing. However, if this problem appears in the multiple-choice section, the student's ability to visualize the graph would be tested, so students would most likely have to draw the graph. Be sure you know how to draw basic curves and find intersections.

EXAMPLE 20: (Calculator Active) Let R be the region bounded by the graphs of $y = \sin\left(\dfrac{\pi x}{4}\right)$ and $y = \cos\left(\dfrac{\pi x}{2}\right) - 1$, shown in the figure. The region R models the surface of a lake. At all points in R at a distance x from the y-axis, the depth of the lake is given by $g(x) = 4x - x^2$. Find the volume of the lake.

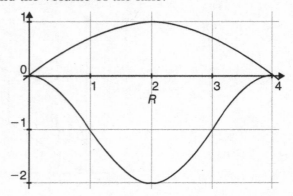

SOLUTION: At any value of x, the volume of a "slice" of water with width dx is the area of the water times its depth which will be

$$\left[\sin\left(\frac{\pi x}{4}\right) - \left(\cos\left(\frac{\pi x}{2}\right) - 1\right)\right] dx \cdot g(x).$$

So $V = \displaystyle\int_0^4 \left[\sin\left(\frac{\pi x}{4}\right) - \left(\cos\left(\frac{\pi x}{2}\right) - 1\right)\right](4x - x^2)\,dx \approx 22.165.$

```
Plot1  Plot2  Plot3
\Y₁= sin(πX/4)
\Y₂= cos(πX/2) − 1
\Y₃=
\Y₄=
\Y₅=
\Y₆=
\Y₇=
```

```
fnInt ((Y₁ − Y₂) (4X
 − X²) , X, 0, 4)
      22.16533736
```

DIDYOUKNOW?

Another Great Pyramid fact: the product of the pyramid's volume and density times 10^{15} is equal to the ratio of the volume and density of the Earth.

Solving Differential Equations

Overview: A differential equation (DEQ) is in the form of $\dfrac{dy}{dx} = \left(\text{Algebraic Expression}\right)$. The goal of solving a DEQ is to work backwards from the derivative to the function; that is to write an equation in the form of $y = f(x) + C$ (since the technique involves integration, the **general solution** will have a constant of integration). If the value of the function at some value of x is known (an initial condition), the value of C can be found. This is called **a specific solution**.

In Calculus AB, the only types of DEQ's studied are called separable. **Separable differential equations** are those than that can be in the form of $f(y)\,dy = g(x)\,dx$. Once in that form, both sides can be integrated with the constant of integration on only one side, usually the right.

Typically there is a problem that requires you to create a **slope field**. Slope fields are a graphical representation of a differential equation by calculating slopes at various points. Simply calculate the slopes using the given derivative formula and plot them with short lines on the given graph. Usually the slopes will be integer values.

DIDYOUKNOW?

Sophus Lie, a nineteenth-century Norwegian mathematician, stated: "Among all of the mathematical disciplines, the theory of differential equations is the most important... It furnishes the explanation of all those elementary manifestations of nature which involve time."

EXAMPLE 21: (Calculator Active) For the differential equation $\dfrac{dy}{dx} = (y-1)^2 e^{-x}$:

    a)   Find the general solution.

    b)   If the initial condition is $\left(0, \dfrac{3}{2}\right)$, find the specific solution of the DEQ.

SOLUTIONS: a) $\dfrac{dy}{(y-1)^2} = e^{-x} dx$

$$\int (y-1)^{-2}\, dy = \int e^{-x}\, dx$$

$$\frac{-1}{y-1} = -e^{-x} + C$$

$$y - 1 = \frac{1}{e^{-x} + C} \Rightarrow y = 1 + \frac{1}{e^{-x} + C}$$

  b) You can go back to any step above to find C.

$$\frac{1}{y-1} = e^{-x} + C$$

$$\frac{1}{\frac{3}{2} - 1} = e^{0} + C$$

$$2 = 1 + C \Rightarrow C = 1$$

$$y = 1 + \frac{1}{e^{-x} + 1}$$

TEST TIP

Just because a graphing calculator is allowed on an AP exam problem doesn't mean that using it will make solving the problem easier. Other than the simple algebra in Example 21, the use of a graphing calculator is of no help. Of the 17 multiple-choice calculator-active questions that have appeared on released AP exams, typically about 7, or 41%, actually required use of a graphing calculator. Although helpful, don't expect that a calculator will solve all of your problems.

EXAMPLE 22: a) On the axes below, sketch a slope field for the given differential equation $\dfrac{dy}{dx} = x^2 y$ at the 16 points indicated and for $-1 < x < 2$, sketch the solution curve through $(-1, 1)$.

b) Find the specific solution of the DEQ passing through the point $(-1, 1)$.

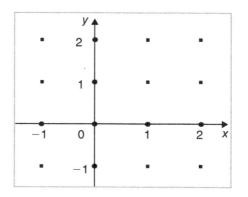

SOLUTIONS: a)

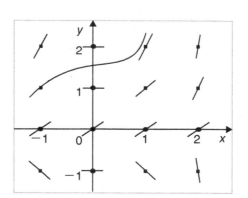

b) $\displaystyle\int \frac{dy}{y} = \int x^2 \, dx$

$\ln|y| = \dfrac{x^3}{3} + C$

$y = e^{\frac{x^3}{3} + C} \;\Rightarrow\; y = Ce^{\frac{x^3}{3}}$

$1 = Ce^{-1/3} \;\Rightarrow\; C = \dfrac{1}{e^{-1/3}} = e^{1/3}$

$y = e^{1/3} e^{x^3/3}$ or $y = e^{\frac{x^3 + 1}{3}}$

EXAMPLE 23: Consider the differential equation $\dfrac{dy}{dx} = x - y + 1$.

    a) On the axes below, sketch the region in the xy-plane in which the solution to the DEQ is concave up.

    b) Let $y = f(x)$ be the particular set of solutions to the differential equation with the condition $f(a) = a + 1$. At these points, does f have a relative maximum, relative minimum, or neither? Justify your answer.

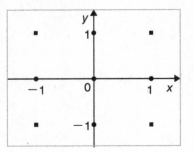

SOLUTIONS: Concavity relates to the second derivative.

    a) $\dfrac{d^2 y}{dx^2} = 1 - \dfrac{dy}{dx} = 1 - (x - y + 1) = y - x$

    Concave up: $\dfrac{d^2 y}{dx^2} > 0$ so $y > x$

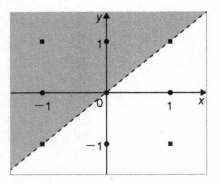

    b) At $(a, a+1)$: $\dfrac{dy}{dx} = a - (a+1) + 1 = 0$

$$\frac{d^2y}{dx^2} = y - x = a + 1 - a = 1 > 0$$

By the 2nd derivative test, $\frac{dy}{dx} = 0$ and $\frac{d^2y}{dx^2} > 0$,

so f has a relative minimum whenever $f(a) = a + 1$.

Growth and Decay

When you encounter word problems involving change with respect to time, they can be set up as a related-rates problem. But when you are given this rate of change of some expression, and wish to find the expression itself, you are usually using models of DEQ's.

A favorite type is a problem using the words: The rate of change of y is proportional to some expression. The equation that describes this statement is: $\frac{dy}{dt} = k \cdot (\text{expression})$ where k is a constant. You use separable differential equation methods to solve this DEQ. Generally, these DEQ's solve into expressions where a quantity **grows** or **decays**.

TEST TIP

Frequently on the AP exam, you are told that the rate of change of y is proportional to y. The DEQ $\frac{dy}{dt} = k \cdot y$ can be solved to the general equation $y = Ce^{kt}$ and it is not necessary to show the calculus involved in creating this equation. These types of problems give exponential growth (when $k > 0$) and exponential decay (when $k < 0$). Exponential growth and decay curves are modeled by the following graphs:

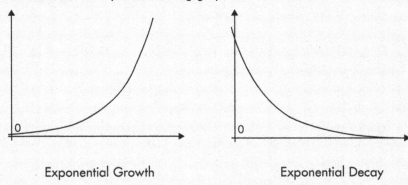

Exponential Growth   Exponential Decay

EXAMPLE 24: (Calculator Active) The rate of change of the height of a sunflower is proportional to its current height. On June 1, a sunflower is 5 inches tall. On June 4, the sunflower is 7 inches tall. How tall will the sunflower be on June 10?

SOLUTION: If h represents the height of the sunflower, the problem states that $\dfrac{dh}{dt} = kh$ and thus $h = Ce^{kt}$. Since we know that $h = 5$ when $t = 0$ (June 1), we concluded that $C = 5$ and $h = 5e^{kt}$. On June 4, $(t = 3)$, we can say: $7 = 5e^{3k}$ and $k = \dfrac{\ln 1.4}{3}$.

Now that we have both C and k, we need to use our formula for h when $t = 9$ (June 10). The sunflower will be 13.72 inches high on June 10.

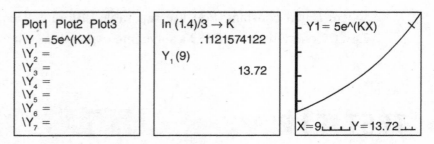

Note that once the differential equation is solved, the problem is no longer a calculus one.

TEST TIP

When variables are used in exponents, just a small change in their value can make a significant difference in expressions that use them. In the AP exam's multiple-choice problems, this fact can lead you to an incorrect solution. For that reason, it is suggested that, rather than rounding values, you store those values on the calculator that you intend to use later. In Example 24, the value of k was stored and thus when it is used later, it will be exact. Any rounding comes only at the very end of the problem.

EXAMPLE 25: A rollercoaster goes up an almost vertical hill and then comes down the other side. When it reaches the bottom of the hill, the coaster is traveling at 120 ft/sec. The track flattens out and the coaster will eventually slow to a stop. On the flat track, the coaster's acceleration is directly proportional to $\dfrac{1}{\sqrt{t}}$, where $t = 0$ is the time when it hits the flat section. The coaster comes to a stop in 25 seconds.

a) Write an expression for the velocity of the coaster.

b) How long is the flat section of the track?

SOLUTIONS: a) $\quad a = \dfrac{dv}{dt} = k\dfrac{1}{\sqrt{t}}$

$$\int dv = \int kt^{-1/2}\, dt$$

$$v = 2kt^{1/2} + C \Rightarrow 120 = 0 + C \Rightarrow C = 120$$

$$v = 2kt^{1/2} + 120$$

$$0 = 2k\sqrt{25} + 120 \Rightarrow k = -12$$

$$v = -24t^{1/2} + 120$$

b) Since velocity > 0, Distance $= \displaystyle\int_0^{25}\left(120 - 24t^{1/2}\right) dt$

$$= \left[120t - 16t^{3/2}\right]_0^{25} = 3000 - 16(125) = 1000 \text{ ft}$$

EXAMPLE 26: Let $A(t)$ represent the number of acres of forest destroyed by a forest fire at time t days where $t \geq 0$. $A(t)$ is increasing at a rate proportional to $950 - A$, where the constant of proportionality is k. At $t = 0$, there have been 400 acres of forest destroyed, and 2 days later, there have been 700 acres destroyed.

a) To the nearest acre, find the amount of forest destroyed by $t = 7$.

b) Find $\lim\limits_{t \to \infty} A(t)$.

SOLUTIONS:

a) $\dfrac{dA}{dt} = k(950 - A)$

$\dfrac{dA}{950 - A} = k \, dt \Rightarrow \displaystyle\int \dfrac{dA}{A - 950} = \int -k \, dt$

$\ln|A - 950| = -kt + C$

$A - 950 = Ce^{-kt} \Rightarrow A = 950 + Ce^{-kt}$

For $t = 0$, $A = 400 \Rightarrow 400 = 950 + C \Rightarrow C = -550$

$A = 950 - 550e^{-kt}$

For $t = 2$, $A = 700 \Rightarrow 700 = 950 - 550e^{-2k}$

$550e^{-2k} = 250 \Rightarrow e^{-2k} = \dfrac{25}{55}$

$-2k = \ln\left(\dfrac{25}{55}\right) \Rightarrow k = \ln\left(\dfrac{5}{11}\right) \Big/ -2 \approx 0.394$

$A = 950 - 550e^{-0.394t}$

$A(7) \approx 915$ acres

b) $\lim\limits_{t \to \infty} A(t) = \lim\limits_{t \to \infty} \left(950 - 550e^{-0.394t}\right) = 950$ acres.

DID**YOU**KNOW?

A quantity that doubles is experiencing exponential growth. Anything that doubles just 10 times is about one thousand times bigger than it was at the beginning. It is said that scientific and technical knowledge is doubling every seven years. That means that as impressive as our knowledge is today, it represents only half of what we will know 7 years from now and just about 12% of what we will know about 20 years from now.

Time for a quiz
- Review strategies in Chapter 2
- Take Quiz 6 at the REA Study Center

(www.rea.com/studycenter)

Take Mini-Test 2
on Chapters 7–8
Go to the REA Study Center
(www.rea.com/studycenter)

AP Calculus BC

Additional Limit and Integration Problems

Overview: AB Calculus began with a study of limits. BC Calculus starts with the same topic, but takes advantage of the student's knowledge of derivatives to find certain types of limits with less work.

In finding $\lim\limits_{x \to a} \dfrac{f(x)}{g(x)}$ using algebra, we first look at $\dfrac{f(a)}{g(a)}$. If this yields $\dfrac{0}{0}$, we attempt to factor $x-a$ from both the numerator and the denominator. If we can do so, we get $\lim\limits_{x \to a} \dfrac{f(x)}{g(x)} = \lim\limits_{x \to a} \dfrac{(x-a)m(x)}{(x-a)n(x)}$. We can then cancel $(x-a)$ because $x-a \neq 0$ as x is approaching a and not equal to a. We then find $\lim\limits_{x \to a} \dfrac{m(x)}{n(x)}$ by calculating $\dfrac{m(a)}{n(a)}$.

L'Hospital's Rule

L'Hospital's Rule simplifies this procedure. It is used with limits if the initial plugging in yields an expression in **indeterminate form**: $\dfrac{0}{0}, \dfrac{\infty}{\infty}, \infty - \infty, 0 \cdot \infty, 1^{\infty}, \infty^{0}$ and 0^{0}. Logic quickly goes down the drain when we work with expressions in these forms. It is possible to construct problems in which a limit in the form of $\infty - \infty$ is 0, ∞, or any number between 0 and ∞.

Simply stated, L'Hospital's Rule states that if $\lim\limits_{x\to a}\dfrac{f(x)}{g(x)}$ produces an indeterminate

form $\dfrac{0}{0}$ or $\dfrac{\infty}{\infty}$, then $\lim\limits_{x\to a}\dfrac{f(x)}{g(x)}=\lim\limits_{x\to a}\dfrac{f'(x)}{g'(x)}$. Note that this is not the quotient rule. You are taking the derivative of the numerator divided by the derivative of the denominator.

L'Hospital's Rule can be used multiple times within a problem. Indeterminate forms $\infty-\infty$ and $0-\infty$ require students to write these expressions as fractions in order that L'Hospital's Rule can be used. Indeterminate forms: $1^{\infty}, \infty^{0}$ and 0^{0} require natural logarithms to be taken of the expression in order to eliminate the exponent and clear the way for L'Hospital's Rule to be used.

EXAMPLE 1: Find $\lim\limits_{x\to 1}\dfrac{x^3-1}{x^2-6x+5}$.

SOLUTION: When we plug in 1, we get $\dfrac{0}{0}$. Doing the problem without L'Hospital's

Rule, we get $\lim\limits_{x\to 1}\dfrac{(x-1)(x^2+x+1)}{(x-1)(x-5)}=\lim\limits_{x\to 1}\dfrac{x^2+x+1}{x-5}=\dfrac{-3}{4}$.

But using L'Hospital's Rule, we get $\lim\limits_{x\to 1}\dfrac{3x^2}{2x-6}=\dfrac{-3}{4}$. Obviously, L'Hospital's Rule is easier, especially if you forget the factorization of x^3-1.

EXAMPLE 2: Find $\lim\limits_{x\to 0}\dfrac{x^4-x^3-2x^2}{\cos x-1}$.

SOLUTION: Plugging in 0, we get $\dfrac{0}{0}$. This problem must be done with L'Hospital's Rule because we cannot factor the denominator.

$\lim\limits_{x\to 0}\dfrac{4x^3-3x^2-4x}{-\sin x}=\dfrac{0}{0}$. We then apply L'Hospital's Rule again and

get $\lim\limits_{x\to 0}\dfrac{12x^2-6x-4}{-\cos x}=\dfrac{-4}{-1}=4$.

EXAMPLE 3: Find $\lim\limits_{x \to \infty} \dfrac{\ln x}{e^x}$.

SOLUTION: Plugging in ∞, we get $\dfrac{\infty}{\infty}$. Applying L'Hospital's Rule, we get

$$\lim\limits_{x \to \infty} \dfrac{\frac{1}{x}}{e^x} = \lim\limits_{x \to \infty} \dfrac{1}{xe^x} = \dfrac{1}{\infty} = 0.$$

TEST TIP

Don't fall in love with L'Hospital's Rule. You must have an indeterminate form after the initial plug-in for L'Hospital's Rule to work. For instance, $\lim\limits_{x \to 0} \dfrac{x^2}{e^x}$ would give 2 if L'Hospital's Rule could be applied, but the true answer by plugging in 0 is 0.

EXAMPLE 4: Find $\lim\limits_{x \to \infty} x \sin \dfrac{1}{x}$.

SOLUTION: Plugging in ∞ we get $\infty \cdot \sin \dfrac{1}{\infty} = \infty \cdot 0$ which is indeterminate. To use L'Hospital's Rule, we have to write $x \sin \dfrac{1}{x}$ as a fraction. We create a complex fraction: $\lim\limits_{x \to \infty} \dfrac{\sin \frac{1}{x}}{\frac{1}{x}} = \dfrac{0}{0}$. Since this is indeterminate, we can apply L'Hospital's Rule: $\lim\limits_{x \to \infty} \dfrac{\sin \frac{1}{x}}{\frac{1}{x}} = \lim\limits_{x \to \infty} \dfrac{\cos \frac{1}{x}\left(\frac{-1}{x^2}\right)}{\frac{-1}{x^2}} = \lim\limits_{x \to \infty} \cos \dfrac{1}{x} = 1.$

EXAMPLE 5: Find $\lim\limits_{x \to 0}\left(\dfrac{1}{x} - \dfrac{1}{e^x - 1}\right)$.

SOLUTION: Plugging in 0, we get $\dfrac{1}{0} - \dfrac{1}{0} = \infty - \infty$. This is indeterminate. To use

L'Hospital's rule, we have to write $\dfrac{1}{x} - \dfrac{1}{e^x - 1}$ as a single fraction.

We do this by getting an LCD: $\dfrac{1}{x} - \dfrac{1}{e^x - 1} = \dfrac{e^x - 1 - x}{x(e^x - 1)}$ and we find

that $\lim\limits_{x \to 0}\left[\dfrac{e^x - 1 - x}{x(e^x - 1)}\right] = \dfrac{1 - 1 - 0}{0(0)} = \dfrac{0}{0}$. Applying L'Hospital's Rule:

$\lim\limits_{x \to 0}\left[\dfrac{e^x - 1 - x}{x(e^x - 1)}\right] = \lim\limits_{x \to 0}\left(\dfrac{e^x - 1}{xe^x + e^x - 1}\right) = \dfrac{0}{0}$. Applying L'Hospital's

Rule one more time we get: $\lim\limits_{x \to 0}\left(\dfrac{e^x}{xe^x + e^x + e^x}\right) = \dfrac{1}{2}$.

EXAMPLE 6: Find $\lim\limits_{x \to \infty} x^{1/x}$.

SOLUTION: Plugging in ∞, we get ∞^0 which is indeterminate. To use L'Hospital's rule, we have to write $x^{1/x}$ as a fraction. Call the expression $x^{1/x} = y$. Take the natural log of both sides. $\ln y = \ln x^{1/x} = \dfrac{\ln x}{x}$. Plugging in ∞, we get $\dfrac{\infty}{\infty}$, opening the door to L'Hospital's Rule: $\lim\limits_{x \to \infty} \dfrac{\frac{1}{x}}{1} = 0$. However, realize that we started the problem by taking the natural log of the expression: L'Hospital's Rule gives a 0 so $\ln y = 0$ and $y = 1$. We conclude that $\lim\limits_{x \to \infty} x^{1/x} = 1$.

TEST TIP

If limit problems occur on AP exam questions that are calculator-active, they can be checked by plugging numbers into the expression. For instance, in Example 6, $\lim_{x \to \infty} x^{1/x}$, use a very large number for x. Realize that using a calculator to justify this limit in a free-response question on the exam would receive no credit. The calculator cannot be used as a justification.

```
9999999^ (1/ 9999999
              1.000001612
```

EXAMPLE 7:   Consider the differential equation $\dfrac{dy}{dx} = x + y$. Let $y = f(x)$ be the particular solution to this differential equation with initial condition $f(2) = 4$. Find $\lim\limits_{x \to 2} \dfrac{x^5 - 8f(x)}{\sin 2\pi x}$.

SOLUTION: $\lim\limits_{x \to 2} \dfrac{x^5 - 8f(x)}{\sin 2\pi x} = \dfrac{2^5 - 8f(2)}{\sin 4\pi} = \dfrac{32 - 8(4)}{0} = \dfrac{0}{0}$. Applying L'Hospital's Rule, we get

$$\lim_{x \to 2} \frac{5x^4 - 8f'(x)}{2\pi \cos 2\pi x} = \lim_{x \to 2} \frac{5x^4 - 8(x + y)}{2\pi \cos 2\pi x} = \lim_{x \to 2} \frac{5(16) - 8(2 + 4)}{2\pi \cos 4\pi} = \frac{80 - 48}{2\pi} = \frac{16}{\pi}.$$

TEST TIP

It can be shown with L'Hospital's Rule that $\lim\limits_{x \to \infty}\left(1 + \dfrac{1}{x}\right)^x = e$. It takes a number of steps to prove this, and it's something that often appears on the AP exam. You should simply memorize it.

DIDYOUKNOW?

Guillaume François Antoine de L'Hospital (1661–1704) was intrigued by calculus and was sent findings by mathematician John Bernoulli, who was paid by L'Hospital. Thus, what we call "L'Hospital's Rule" was first discovered by Bernoulli.

Integration by Parts

Overview: While the derivative of any expression can be taken, the same is not true for indefinite integrals. BC Calculus increases students' ability to integrate functions. We start with **integration by parts** which can be applied to a variety of functions and is particularly suited for integrands involving products of expressions.

The basic rule for integration by parts states that if u and v are functions of x with continuous derivatives, then $\int u\,dv = uv - \int v\,du$. Integration by parts trades one integral for another that might be easier to take. Repeated use of integration by parts may be necessary within a problem.

Students should not automatically turn to integration by parts when attacking an integration problem. Conventional u-substitution should always be the first resort when encountering such a problem, as "parts" can make the problem more difficult.

The best way to start integration by parts is to set up a chart:

$$u = \qquad v =$$
$$du = \qquad dv =$$

You need to take all components of the integrand and distribute them to u and dv. The dx will always go with the dv. There is no specific rule concerning how to divide up the other components but, in general, functions that can be "powered down" are typically the u and functions that have repetitive derivatives (exponential and trig) are typically the dv. Many times, though, it is simply trial-and-error and experience that solve these types of problems.

EXAMPLE 8: Find $\int 2xe^x \, dx$.

SOLUTION: Your first thought should be u-substitution. But if $u = x$ (the exponent of the e), then $du = dx$ and there is nothing we can do with $2x$. So we have to think integration by parts.

Since e^x is easy to integrate, here is a way to divide up our components:

$$u = 2x \qquad v =$$
$$du = \qquad dv = e^x dx$$

At this point, take the derivative of u and take the integral of dv:

$$u = 2x \qquad v = e^x$$
$$du = 2 \, dx \qquad dv = e^x dx$$

Now apply the integration by parts formula:

$\int 2xe^x \, dx = 2xe^x - \int 2e^x \, dx$. We see that the resulting integral is easy to take and thus $\int 2xe^x \, dx = 2xe^x - 2e^x + C$.

This solution can be verified by taking the derivative of the solution: $\dfrac{d}{dx}\left(2xe^x - 2e^x + C\right) = 2xe^x + 2e^x - 2e^x = 2xe^x$.

TEST TIP

You are more likely to encounter problems emphasizing integration techniques in the free-response section of the AP exam rather than in the multiple-choice section. The reason is that it is possible to use trial-and-error to take the derivatives of all answers to find which gives the expression to be integrated. If you do encounter integration problems in the multiple-choice section, use this trial-and-error approach as a last resort as it can be time-consuming.

EXAMPLE 9: Find $\int x^4 \ln x \, dx$.

SOLUTION: $u = \ln x \qquad v = \dfrac{x^5}{5}$

$du = \dfrac{1}{x} \, dx \qquad dv = x^4 dx$

$\int x^4 \ln x \, dx = \dfrac{x^5}{5} \ln x - \int \left(\dfrac{x^5}{5} \right) \left(\dfrac{1}{x} \right) dx = \dfrac{x^5}{5} \ln x - \dfrac{x^5}{25} + C$

EXAMPLE 10: Find $\int x \sin x \, dx$.

SOLUTION: $u = x \qquad v = -\cos x$

$du = dx \qquad dv = \sin x \, dx$

$\int x \sin x \, dx = -x \cos x - \int (-\cos x) \, dx = -x \cos x + \sin x + C$

EXAMPLE 11: Find the region between the x-axis, the curve $y = \ln x$, and the line $x = e^2$.

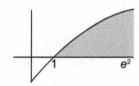

SOLUTION: $A = \int\limits_{1}^{e^2} \ln x \, dx$

$u = \ln x \qquad v = x$

$du = \dfrac{1}{x} dx \qquad dv = dx$

$\int \ln x \, dx = x \ln x - \int (x) \left(\dfrac{1}{x} \right) dx = x \ln x - x$

$\int\limits_{1}^{e^2} \ln x \, dx = [x \ln x - x]_{1}^{e^2} = e^2 \ln e^2 - e^2 - (\ln 1 - 1) = 2e^2 - e^2 + 1 = e^2 + 1$

TEST TIP

Example 11 involving $\int \ln x\, dx$ usually finds its way into the AP exam. Some teachers recommend memorizing the solution $(x \ln x - x + C)$ rather than go through integration by parts.

EXAMPLE 12: Find $\int x^2 e^{-x} dx$.

SOLUTION: In this problem, integration by parts must be used twice:

$$\int x^2 e^{-x}\, dx = -x^2 e^{-x} - \int 2x(-e^{-x})\, dx \qquad u = x^2 \qquad v = -e^{-x}$$

$$= -x^2 e^{-x} + 2\int x e^{-x}\, dx \qquad\qquad \underline{du = 2x\, dx \quad dv = e^{-x} dx}$$

$$= -x^2 e^{-x} + 2\left[x(-e^{-x}) - \int -e^{-x}\, dx \right] \qquad u = x \qquad v = -e^{-x}$$

$$= -x^2 e^{-x} - 2x e^{-x} - 2e^{-x} + C \qquad\qquad du = 1\, dx \quad dv = e^{-x} dx$$

$$= -e^{-x}(x^2 + 2x + 2) + C$$

EXAMPLE 13: Find $\int x e^{x^2}\, dx$.

SOLUTION: This is a trap problem. Integration by parts doesn't work and worse, wastes time. This is a standard u-substitution problem:

$$\int x e^{x^2}\, dx = \frac{1}{2}\int 2x e^{x^2}\, dx \qquad u = x^2 \qquad du = 2x\, dx$$

$$= \frac{1}{2}\int e^u\, du = \frac{1}{2}e^{x^2} + C$$

EXAMPLE 14: A particle moves along the x-axis so that at any time $t > 0$, its velocity is given by $v(t) = t^2 \sin t$. At time $t = 0$, the position of the particle is $x(0) = 3$. Find the position function $x(t)$ of the particle.

SOLUTION:

$$x(t) = \int t^2 \sin t \; dt$$

$$= -t^2 \cos t + \int 2t \cos t \; dt \qquad\qquad \begin{array}{ll} u = t^2 & v = -\cos t \\ du = 2t \; dt & dv = \sin t \; dt \\ u = 2t & v = \sin t \\ du = 2 \; dt & dv = \cos t \; dt \end{array}$$

$$= -t^2 \cos t + \left(2t \sin t - \int 2 \sin t \; dt \right)$$

$$= -t^2 \cos t + 2t \sin t + 2\cos t + C$$

$$x(0) = 0 + 0 + 2 + C = 3 \Rightarrow C = 1$$

$$x(t) = -t^2 \cos t + 2t \sin t + 2\cos t + 1$$

Integration Using Partial Fractions

Overview: When you attempt to integrate a fraction, typically you let u be the expression in the denominator and hope that du will be in the numerator. When this doesn't happen, the technique of **partial fraction decomposition** may work. One form of this type of problem is $\int \dfrac{mx+n}{(x-a)(x-b)} \, dx$ where $a \neq b$.

You need to write $\dfrac{mx+n}{(x-a)(x-b)}$ as the sum of two fractions: $\dfrac{}{x-a} + \dfrac{}{x-b}$. To find the numerator of the $x-a$ expression, cover up the $x-a$ in $\dfrac{mx+n}{(x-a)(x-b)}$ expression, and plug in $x = a$. To find the numerator of the $x-b$ expression, cover up the $x-b$ in $\dfrac{mx+n}{(x-a)(x-b)}$ expression, and plug in $x = b$. From there, each expression can be integrated using a natural log expression.

EXAMPLE 15: Find $\int \dfrac{6x-2}{x^2-2x-3}\,dx$.

SOLUTION: $\int \dfrac{6x-2}{x^2-2x-3}\,dx = \int \dfrac{6x-2}{(x-3)(x+1)}\,dx$

$$\dfrac{6x-2}{(x-3)(x+1)} = \dfrac{\frac{6(3)-2}{3+1}}{x-3} + \dfrac{\frac{6(-1)-2}{-1-3}}{x+1} = \dfrac{4}{x-3} + \dfrac{2}{x+1}$$

$$\int\left[\dfrac{4}{x-3} + \dfrac{2}{x+1}\right] dx = 4\ln|x-3| + 2\ln|x+1| + C$$

TEST TIP

Students need to be aware of logarithm rules when encountering these problems on the multiple-choice section of the AP exam. In Example 15, 4 ln|x − 3|+2 ln|x + 1| can be written as 2 ln|x − 3|² +2 ln|x + 1| and further transformed into 2 [ln|(x − 3)² (x + 1)|] + C. While the original answer is fine as a free-response answer, these transformations can easily trap a student in the multiple-choice section.

EXAMPLE 16: Find $\int \dfrac{x^2+1}{x^3-16x}\,dx$.

SOLUTION:

$$\int \dfrac{x^2+1}{x^3-16x}\,dx = \int \dfrac{x^2+1}{x(x-4)(x+4)}\,dx$$

$$\dfrac{x^2+1}{x(x-4)(x+4)} = \dfrac{\frac{-1}{16}}{x} + \dfrac{\frac{17}{32}}{x-4} + \dfrac{\frac{17}{32}}{x+4}$$

$$\int\left[\dfrac{-\frac{1}{16}}{x} + \dfrac{\frac{17}{32}}{x-4} + \dfrac{\frac{17}{32}}{x+4}\right] dx = \dfrac{17}{32}\ln|x-4| + \dfrac{17}{32}\ln|x+4| - \dfrac{1}{16}\ln|x| + C$$

This could also be written as: $\dfrac{1}{32}\ln\left(\dfrac{|x-4|^{17}|x+4|^{17}}{x^2}\right) + C$

EXAMPLE 17: Find $\int \dfrac{x^3}{x^2-1}\,dx$.

SOLUTION: To use this method, the degree of the numerator must be greater than the degree of the denominator. Do long division.

$$\int \frac{x^3}{x^2-1}\,dx = \int \left(x + \frac{x}{x^2-1} \right) dx$$

$$x^2-1\overline{\smash{\big)}\,x^3}$$
$$\underline{x^3 - x}$$
$$x$$

$$x + \frac{x}{x^2-1} = x + \frac{\frac{1}{2}}{x-1} + \frac{\frac{1}{2}}{x+1}$$

$$\int \left[x + \frac{\frac{1}{2}}{x-1} + \frac{\frac{1}{2}}{x+1} \right] dx = \frac{x^2}{2} + \frac{1}{2}\ln|x-1| + \frac{1}{2}\ln|x+1|$$

$$= \frac{x^2 + \ln|x^2-1|}{2} + C$$

Note that $\int \dfrac{x}{x^2-1}\,dx$ can be solved by integration by parts or by u-substitution.

EXAMPLE 18: A window is modeled in the figure below using the following equations as borders: $y = \dfrac{1}{x^2-8x+7}$, $y = -1$, $x = 2$, $x = 6$. Find the area of the window:

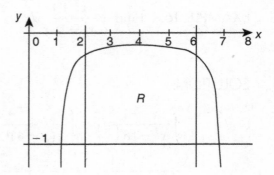

$$A = \int_{2}^{6} \left[\frac{1}{x^2-8x+7} - (-1) \right] dx$$

$$A = \int_{2}^{6} \left[\frac{1}{(x-7)(x-1)} + 1 \right] dx$$

$$A = \int_{2}^{6} \left[\frac{\frac{1}{6}}{x-7} - \frac{\frac{1}{6}}{x-1} + 1 \right] dx$$

$$A = \left[\frac{1}{6} \ln \left| \frac{x-7}{x-1} \right| + x \right]_2^6$$

$$A = \frac{1}{6} \ln \frac{1}{5} + 6 - \left(\frac{1}{6} \ln 5 + 2 \right) = \frac{1}{6} \left(\ln \frac{1}{5} - \ln 5 \right) + 4$$

This could be written as: $\frac{1}{6} \ln \left(\frac{1}{25} \right) + 4$ or $4 - \frac{\ln 25}{6}$ or $4 - \frac{\ln 5}{3}$

DIDYOU**KNOW?**

The method just described in performing partial fraction decomposition is called the Heaviside Cover-Up Method named after mathematician Oliver Heaviside (1850–1925), a self-taught English mathematician, engineer, and physicist who changed the face of mathematics for many years.

Improper Integrals

Overview: An **improper integral** is in the form $\int_b^\infty f(x)\, dx$ or $\int_a^a f(x)\, dx$ or $\int_{-\infty}^\infty f(x)\, dx$. It also can be in the form $\int_a^b f(x)\, dx$ where there is at least one value c such that $a \le c \le b$ for which $f(x)$ is not continuous. Improper integrals usually go hand-in-hand with area and volume problems and typically integration by parts might be needed to integrate such expressions.

Improper integrals are just limit problems in disguise: $\int_a^\infty f(x)\, dx = \lim_{b \to \infty} \int_a^b f(x)\, dx$ or $\int_{-\infty}^b f(x)\, dx = \lim_{a \to -\infty} \int_a^b f(x)\, dx$. Although x is getting infinitely larger, the area under a function $f(x)$ may also be getting larger, but not infinitely larger. If this area approaches a limit, the improper integral is said to **converge** to this limit. Otherwise, if the area gets infinitely large, the improper integral is said to **diverge**.

In the case where there is a discontinuity at $x = c$, the improper integral is split into two pieces: $\int_a^b f(x)\, dx = \lim_{k \to c^-} \int_a^k f(x)\, dx + \lim_{k \to c^+} \int_k^b f(x)\, dx$. If either of these integrals diverges, the improper integral diverges.

EXAMPLE 19: Determine if $\int\limits_0^\infty \dfrac{10}{x+10}\,dx$ converges or diverges.

SOLUTION: $\int\limits_0^\infty \dfrac{10}{x+10}\,dx = 10\ln|x+10|\,\big|_0^\infty = 10\ln(\infty) - 10\ln 10$ which is divergent.

TEST TIP

Technically, the correct notation for Example 19 is $\lim\limits_{b\to\infty}\left(10\ln|x+10|\,\big|_0^b\right)$. If this type of problem occurs on the free-response section of the AP exam, the shorter notation provided in the solution of Example 19 is acceptable.

EXAMPLE 20: Determine if $\int\limits_1^\infty \dfrac{1}{x^4}\,dx$ is convergent and, if so, find what it converges to.

SOLUTION: $\int\limits_1^\infty \dfrac{1}{x^4}\,dx = \int\limits_1^\infty x^{-4}\,dx = \left[\dfrac{x^{-3}}{-3}\right]_1^\infty = \left[\dfrac{1}{-3x^3}\right]_1^\infty = 0 - \left(\dfrac{-1}{3}\right) = \dfrac{1}{3}$

EXAMPLE 21: Determine if $\int\limits_1^\infty \dfrac{1}{\sqrt{x}}\,dx$ is convergent and, if so, find what it converges to.

SOLUTION: $\int\limits_1^\infty \dfrac{1}{\sqrt{x}}\,dx = \int\limits_1^\infty x^{-1/2}\,dx = \left[\dfrac{x^{1/2}}{1/2}\right]_1^\infty = \left[2\sqrt{x}\right]_1^\infty = 2\sqrt{\infty} - 2$

which is divergent.

EXAMPLE 22: Determine if $\int\limits_0^\infty \dfrac{5x}{e^x}\,dx$ is convergent and, if so, find what it converges to.

SOLUTION:
$$\int_0^\infty \frac{5x}{e^x} dx = \int_0^\infty (5x)e^{-x} dx$$

$$\left[-5xe^{-x}\right]_0^\infty - \left[\int_0^\infty -5e^{-x} dx\right]$$

$u = 5x \qquad v = -e^{-x}$

$du = 5\,dx \quad dv = e^{-x}dx$

$$\left[\frac{-5x}{e^x}\right]_0^\infty - \left[5e^{-x}\right]_0^\infty$$

$$\lim_{x\to\infty}\left[\frac{-5}{e^x}\right] - \left(\frac{0}{1}\right) - \left[\frac{5}{e^x}\right]_0^\infty \qquad \text{L'Hospital's Rule}$$

$$0-(0)-(0-5)=5$$

EXAMPLE 23: Determine if $\int_{-4}^4 \frac{1}{x} dx$ is convergent and, if so, find what it converges to.

SOLUTION: Since $y = \frac{1}{x}$ is not continuous at $x = 0$, this is an improper integral and must be divided into two integrals.

$$\int_{-4}^4 \frac{1}{x} dx = \int_{-4}^0 \frac{1}{x} dx + \int_0^4 \frac{1}{x} dx$$
$$= \ln|x|_{-4}^0 + \ln|x|_0^4$$
$$= \ln 0 - \ln 4 + \ln 4 - \ln 0$$

This is divergent despite the temptation to cancel out the ln 0 expressions. Since ln 0 does not exist, we cannot do arithmetic with this expression.

Students have difficulty with this concept, as they may argue that the area under the x-axis between -4 and 0 is the same as the area above the x-axis between 0 and 4 (and thus cancels out). Then the answer would be zero. But these areas are infinite, and we have already mentioned that $\infty - \infty$ does not necessarily equal zero.

EXAMPLE 24: Find the unbounded area between the graph of $y = \dfrac{100}{x^2+1}$ and the x-axis.

SOLUTION:
$$A = \int_{-\infty}^\infty \frac{100}{x^2+1} dx$$

$$A = 100\left[\tan^{-1} x\right]_{-\infty}^\infty = 100\tan^{-1}(\infty) - 100\tan^{-1}(-\infty)$$

$$A = 100\left(\frac{\pi}{2}\right) - 100\left(-\frac{\pi}{2}\right) = 100\pi$$

EXAMPLE 25: The shaded region in the figure below is between the line $x = 1$, the graph of $y = \dfrac{1}{x^4}$ and the x-axis. Find the value of k for which the line $x = k$ divides this region into two equal parts.

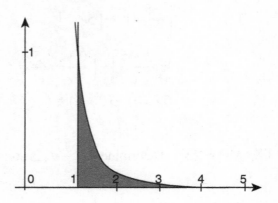

SOLUTION:

$$\int_1^k \frac{1}{x^4}dx = \int_k^\infty \frac{1}{x^4}dx$$

$$-\frac{1}{3x^3}\Big|_1^k = -\frac{1}{3x^3}\Big|_k^\infty$$

$$-\frac{1}{3k^3} + \frac{1}{3} = 0 + \frac{1}{3k^3}$$

$$\frac{2}{3k^3} = \frac{1}{3} \Rightarrow k = \sqrt[3]{2}$$

OR

$$\int_1^k \frac{1}{x^4}dx = \frac{1}{2}\int_1^\infty \frac{1}{x^4}dx$$

$$-2\left[\frac{1}{3x^3}\right]_1^k = -\frac{1}{3x^3}\Big|_1^\infty$$

$$-\frac{2}{3k^3} + \frac{2}{3} = 0 + \frac{1}{3}$$

$$\frac{2}{3k^3} = \frac{1}{3} \Rightarrow k = \sqrt[3]{2}$$

EXAMPLE 26: Find the volume of the solid formed by rotating the unbounded first quadrant region lying between the graph of $y = e^{-x}$ and the x-axis about the x-axis.

SOLUTION:

$$V = \pi\int_0^\infty \left(e^{-x}\right)^2 dx = \pi\int_0^\infty e^{-2x}dx$$

$$V = \pi\left[\frac{-1}{2}e^{-2x}\right]_0^\infty$$

$$V = \pi\left[\frac{-1}{2e^{2x}}\right]_0^\infty$$

$$V = \pi\left[0 - \left(-\frac{1}{2}\right)\right] = \frac{\pi}{2}$$

TEST TIP

Confirming the result of Example 26 is dangerous using the calculator. Note that plugging in 999 as our large number gives the result from Example 26: $\frac{1}{2} = 0.5$. But using a bigger number (99999) gives an answer of zero. The reason is the method the calculator uses to numerically calculate the integral divides the area into such small sections that are less than the calculator's accuracy level of 10 decimal places. It is essentially adding a bunch of zeros.

```
fnInt ((e^(-X))², X, 0, 999
                          .5
fnInt((e^(-X))², X, 0, 99999
                           0
```

Time for a quiz
- Review strategies in Chapter 2
- Take Quiz 7 at the REA Study Center
  (www.rea.com/studycenter)

Additional BC Calculus Applications

Euler's Method

Overview: Euler's Method provides a numerical procedure to approximate the solution of a differential equation with a given initial value. Typically, you are given a differential equation $\dfrac{dy}{dx}$ and a point on the function (x_0, y_0). Your goal is to approximate the value of the function at a value of $x_0 + n\Delta x$ where Δx is given and n is an integer. Euler's Method is a **recursive procedure**: you start with a number, run through the procedure, and use the resulting answer as the input of the same procedure. Each time you run through the procedure, it is called an **iteration**.

The meanings of Δx, dx, Δy and dy can be seen in the following figure.

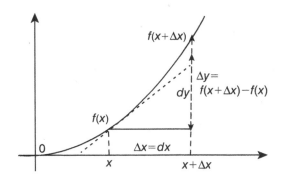

In a continuous function f, we choose some value x and draw the tangent line to f at x with slope $\dfrac{dy}{dx}$. We then move a distance Δx or dx to the right and find $f(x + \Delta x)$. $\Delta y = f(x + \Delta x) - f(x)$ which is the vertical dashed line in the figure. dy, the angled dashed line, is an approximation for Δy. We are usually interested in Δy, the actually change in y, but if Δx is close to zero, $\Delta y \approx dy$. Euler's Method uses this value of $dy \approx \dfrac{dy}{dx}\Delta x$ to move from point to point on the function. Since we are using approximations, the more iterations of Euler's Method that are applied, usually the further off our values of y are. So, Euler's Method uses small values of Δx and few steps. Typically Euler's Method can involve messy decimals after a few steps. For that reason, Euler's Method problems usually occur in the non-calculator section of the AP exam where only a few steps of the procedure need to be performed.

Procedure:

1) Start with a differential equation (DEQ), a given initial point (x_0, y_0) on the graph of the function and a given $\Delta x = dx$.

2) Calculate the slope $\dfrac{dy}{dx}$ using the DEQ at the point.

3) Approximate the value of dy using the fact that $dy \approx \dfrac{dy}{dx}\Delta x$.

4) Find the new values of y and x: $y_{new} = y_{old} + dy$ and $x_{new} = x_{old} + \Delta x$.

5) Repeat the process at step 2).

EXAMPLE 1: Let $y = f(x)$ be the solution to the differential equation $\dfrac{dy}{dx} = x^2 - 2y$ with the initial condition that $f(2) = -1$. What is the approximation for $f(3)$ if Euler's Method is used, starting at 2 with a step size of 0.5?

SOLUTION: It is best to make a chart in an Euler's Method problem.

x	2	2.5	3
$y_{new} = y_{old} + \dfrac{dy}{dx}\Delta x$	-1	$-1 + 6(0.5) = 2$	$2 + 2.25(0.5) = 3.125$
$\dfrac{dy}{dx}$	$2^2 - 2(-1) = 6$	$2.5^2 - 2(2) = 2.25$	

Thus, $f(3) \approx 3.125$.

EXAMPLE 2:   (Calculator Active) Consider the differential equation $\dfrac{dy}{dx} = -6x^2 y$ with initial condition $f(0) = 2$. Find the difference between the exact value of $f(1)$ and an Euler approximation of $f(1)$ using two equal steps.

SOLUTION:

Euler Approximation

x	0	0.5	1
$y_{new} = y_{old} + \dfrac{dy}{dx} \Delta x$	2	$2 + 0(0.5) = 2$	$2 + (-3)(.5) = 0.5$
$\dfrac{dy}{dx}$	$-6(0)(2) = 0$	$-6(5^2)(2) = -3$	

Exact value: $\dfrac{dy}{dx} = -6x^2 y \Rightarrow \dfrac{dy}{y} = -6x^2 dx$

$$\int \frac{dy}{y} = \int -6x^2\, dx$$

$$\ln|y| = -2x^3 + C \Rightarrow y = Ce^{-2x^3}$$

$$\text{At } x = 0:\ 2 = Ce^0 \Rightarrow C = 2 \Rightarrow y = 2e^{-2x^3}$$

$$\text{At } x = 1:\ y = 2e^{-2} = \frac{2}{e^2}$$

$$\text{Difference: } 0.5 - \frac{2}{e^2} \approx 0.229$$

DIDYOUKNOW?

Euler's Method was named after Leonhard Euler (pronounced "Oiler") (1707–1783). A Swiss mathematician and physicist, Euler is considered to be one of the greatest mathematicians to have ever lived. Euler introduced the concept of a function and was the first to use the notation $f(x)$. He introduced the modern notation for trigonometric functions, the letter e for the base of the natural numbers (known as Euler's number), and the letter i to denote the imaginary unit. He also popularized the Greek letter π to denote the ratio of a circle's circumference to its diameter.

Logistic Growth

Overview: Logistic curves occur when a quantity is growing at a rate proportional to itself and the room available for growth. This room available is called the carrying capacity. This constantly increasing curve, as shown in the figure below, has a distinctive "S-shape" where the initial stage of growth is exponential, then slows, and eventually the growth essentially stops, when the curve approaches the **carrying capacity**.

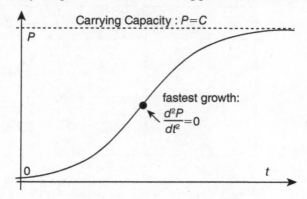

Logistic growth is signaled by the differential equation $\dfrac{dP}{dt} = kP(C - P)$. While this DEQ can be solved as $P(t) = \dfrac{C}{1 + de^{-Ckt}}$, where C, d, and k are constants, no released AP questions have asked students to know that equation. Rather they need to know how to recognize a logistic growth situation and determine the population when the logistic growth is the fastest. This is accomplished by $\dfrac{d^2P}{dt^2} = 0$. Also students need to know that the curve has a horizontal asymptote meaning $\lim\limits_{t \to \infty} \dfrac{dP}{dt} = 0$ and thus $\lim\limits_{t \to \infty} P(t) = C$ (the carrying capacity).

EXAMPLE 3: A population is modeled by a function P that satisfies the logistic differential equation

$$\frac{dP}{dt} = \frac{P}{10}\left(1 - \frac{P}{20}\right)$$

a) If $P(0) = 5$, find $\lim\limits_{t \to \infty} P(t)$.

b) If $P(0) = 5$, for what value of P is the population growing the fastest?

SOLUTIONS: Since this is logistic growth, there is a horizontal asymptote to $P(t)$ and thus $\lim\limits_{t\to\infty}\dfrac{dP}{dt}=0$.

a) $\lim\limits_{t\to\infty}\dfrac{dP}{dt}=\lim\limits_{t\to\infty}\dfrac{P}{10}\left(1-\dfrac{P}{20}\right)=0$

$1-\dfrac{P}{20}=0\Rightarrow\lim\limits_{t\to\infty}P=20$

b) $\dfrac{dP}{dt}=\dfrac{P}{10}\left(1-\dfrac{P}{20}\right)=\dfrac{P}{10}-\dfrac{P^2}{200}$

$\dfrac{d^2P}{dt^2}=\left(\dfrac{1}{10}-\dfrac{P}{100}\right)\dfrac{dP}{dt}=0$

Since $\dfrac{dP}{dt}>0$, $10P=100\Rightarrow P=10$ $\left(\text{half the carrying capacity}\right)$

EXAMPLE 4: Let g be a function with $g(0)=2$ such that all points (x,y) on the graph of g satisfy the logistic differential equation: $\dfrac{dy}{dx}=8y\left(8-\dfrac{y}{3}\right)$.

a) For what value of y does the graph of g have a point of inflection? Justify your answer.

b) Find the slope of the graph of g at the point of inflection.

SOLUTIONS: a) $\lim\limits_{x\to\infty}\dfrac{dy}{dx}=0\Rightarrow 8-\dfrac{y}{3}=0\Rightarrow y=24$ (carrying capacity)

$\dfrac{d^2y}{dx^2}=\left(64-\dfrac{16y}{3}\right)\dfrac{dy}{dx}=0$

Inflection point: $64-\dfrac{16y}{3}=0\Rightarrow y=12$.

(half the carrying capacity)

b) $\dfrac{dy}{dx}=8(12)\left(8-\dfrac{12}{3}\right)=384$.

An application of logistic functions, given its name in 1845 by Pierre Verhulst, is the spread of rumors. Rumors involve two types of people: tellers and hearers. At first, there are very few tellers so the growth of the rumor is fast as there is a lot of room for growth. But over time, hearers become tellers, and the room for growth is small. So the spread of the rumor slows. Eventually, when there are no more hearers, everyone knows the rumor and the spread stops.

Arc Length

Overview: Given a function on an interval $[a, b]$, the **arc length** is defined as the total length of the function from $x = a$ to $x = b$. For this section, we will only concentrate on curves that are defined in function form. Functions defined parametrically, in vector-valued form, or in polar form have their own formulas derived from the formula below.

Using the figure below and the Pythagorean Theorem, a curve between two points in the xy-plane has length $L = \lim\limits_{\Delta x \to 0, \Delta y \to 0} \sqrt{\Delta x^2 + \Delta y^2}$. This says that when Δx is small, the length of the curve between the points $(x, f(x))$ and $(x + \Delta x, f(x + \Delta x))$ is very close to the line connecting those points. Algebraically it can be shown this is equivalent to saying the arc length of a continuous function $f(x)$ over an interval $[a, b]$ is given by

$L = \int_a^b \sqrt{1 + [f'(x)]^2}\, dx$. The formula can also be written in terms of integrating in terms

of y: $L = \int_c^d \sqrt{1 + [f'(y)]^2}\, dy$. Most problems involving arc length need calculators because

of the difficulty of integrating the expression.

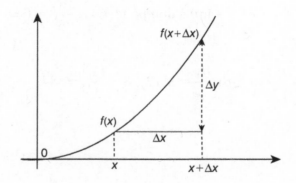

EXAMPLE 5: Find the arc length of the graph of $y = \frac{2}{3}(x^2+1)^{3/2}$ for $0 \le x \le 2$.

SOLUTION: $y' = (x^2+1)^{1/2}(2x) = 2x(x^2+1)^{1/2}$

$$L = \int_0^2 \sqrt{1+4x^2(x^2+1)}\ dx = \int_0^2 \sqrt{4x^4+4x^2+1}\ dx$$

$$L = \int_0^2 \sqrt{(2x^2+1)^2}\ dx$$

$$= \int_0^2 (2x^2+1)\ dx$$

$$L = \left[\frac{2}{3}x^3 + x\right]_0^2 = \frac{16}{3} + 2 = \frac{22}{3}$$

EXAMPLE 6: (Calculator Active) The boundaries of a park is the region R bounded by the graphs of $y = 3x$, $y = 8 - \frac{x^2}{2}$, and the x–axis as shown in the figure below. Find the perimeter of the park, L.

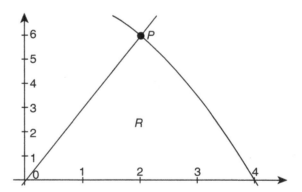

SOLUTION: Point P represents the intersection of the graphs of $y = 3x$ and $y = 8 - \frac{x^2}{2}$.

Then $3x = 8 - \frac{x^2}{2} \Rightarrow x^2 + 6x - 16 = 0 \Rightarrow (x+8)(x-2) = 0$. Reject $x = -8$. Since $x = 2$, $y = 8 - \frac{4}{2} = 6$. Thus, point P is $(2, 6)$.

The distance between $(0, 0)$ and $(4, 0)$ is 4 and from $(0, 0)$ to $(2, 6)$ is $\sqrt{40} = 2\sqrt{10}$.

$L = 4 + 2\sqrt{10} + \int_2^4 \sqrt{1 + \left[f'(x)\right]^2}\, dx$ where the integral represents the length of the curve between $(2, 6)$ and $(4, 0)$.

$$P = 4 + 2\sqrt{10} + \int_2^4 \sqrt{1+(-x)^2}\,dx = 4 + 2\sqrt{10} + \int_2^4 \sqrt{1+x^2}\,dx$$

$$P = 4 + 2\sqrt{10} + 6.336 \approx 16.660$$

EXAMPLE 7: (Calculator Active) The first portion of a roller coaster track is modeled in the figure below. It starts with a lift hill which is to the left of the y-axis. It then transitions into the drop curve $y = x^3 - 5x^2 + 3x + 9$ for $0 \le x \le 3$. At the transition point, the lift hill is tangent to the drop curve. Find the length of the track containing the lift hill and drop curve.

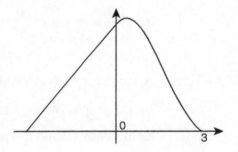

SOLUTION: $y' = 3x^2 - 10x + 3$, $y'(0) = 3$, $y(0) = 9$

Lift hill: $y = 3x + 9$

Lift hill is a line passing through $(0, 9)$ and $(-3, 0)$. Its length $= \sqrt{81+9} = 3\sqrt{10}$

Length of drop curve: $\int_0^3 \sqrt{1+\left(3x^2 - 10x + 3\right)^2}\,dx \approx 10.618$

Track length: $3\sqrt{10} + 10.618 \approx 20.105$

EXAMPLE 8: (Calculator Active) A hill is in the shape of $y = x \cos x$ as shown in the graph below. There is a tunnel along the x-axis that bores through the hill. If a hiker were to walk through the tunnel rather than hiking up and then down the hill, how much distance would he save?

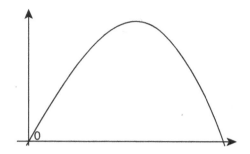

SOLUTION: $y' = -x\sin x + \cos x$

Length of hill: $L = \int\limits_{0}^{\pi/2} \sqrt{1 + (\cos x - x\sin x)^2}\,dx \approx 1.996$

Length of tunnel: $x\cos x = 0 \Rightarrow x = \dfrac{\pi}{2}$

Distance saved: $1.996 - \dfrac{\pi}{2} \approx 0.425$

DIDYOUKNOW?

On a unit circle (radius = 1) in the first quadrant, choose an arc of any length s. No matter where s is placed, the sum of the areas between s and the x-axis and area between s and the y-axis is equal to s. In the figure below, note that regions A and B overlap and the overlap is counted twice. To test yourself on your knowledge of area and arc length for the AP exam, choose two x-values between 0 and 1 and confirm that the two areas add up to the arc length.

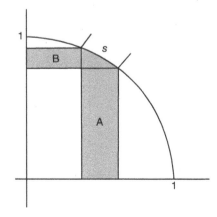

Parametric Functions

Overview: Parametric equations are continuous functions of t in the form $x = f(t)$ and $y = g(t)$. Taken together, the parametric equations create a graph where the points x and y are independent of each other and both dependent on the parameter t. When graphed, parametric curves do not have to be functions. Typically, it is necessary to take derivatives of parametrics. Since the study of vectors parallels the study of parametrics, in this section we will only analyze problems that are not associated with motion in the plane.

If a smooth curve C is given by the parametric equations $x = f(t)$ and $y = g(t)$, then the slope of C at the point (x, y) is given by $\dfrac{dy}{dx} = \dfrac{dy/dt}{dx/dt}$, $dx/dt \neq 0$.

The 2nd derivative of the curve is given by $\dfrac{d^2 y}{dx^2} = \dfrac{d}{dx}\left(\dfrac{dy}{dx}\right) = \dfrac{\dfrac{d}{dt}\left(\dfrac{dy}{dx}\right)}{\dfrac{dx}{dt}}$.

The arc length between $t = a$ and $t = b$ is given by

$$L = \int_{t=a}^{t=b} \sqrt{\left(\dfrac{dx}{dt}\right)^2 + \left(\dfrac{dy}{dt}\right)^2}\, dt.$$ The curve must be smooth and may not intersect itself.

EXAMPLE 9: A curve C is defined by the parametric equations $x = t^3 - t^2 + 1$ and $y = t^2 - 5t - 2$. Find the equation of the line tangent to the graph of C at the point $(5, -8)$.

SOLUTION:

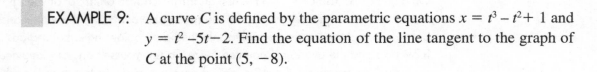

$x = t^3 - t^2 + 1 = 5 \Rightarrow t^3 - t^2 - 4 = 0 \Rightarrow (t-2)(t^2 + t + 2) = 0 \Rightarrow t = 2, \dfrac{-1 \pm i\sqrt{7}}{2}$

$y = t^2 - 5t - 2 = -8 \Rightarrow t^2 - 5t + 6 = 0 \Rightarrow (t-2)(t-3) = 0 \Rightarrow t = 2, 3$

Only $t = 2$ satisfies both equations.

$\dfrac{dy}{dx}_{(5,-8)} = \dfrac{\dfrac{dy}{dt}}{\dfrac{dx}{dt}} = \dfrac{2t-5}{3t^2 - 2t} \Rightarrow \dfrac{dy}{dx}_{[t=2]} = \dfrac{-1}{8}$

Tangent line: $y + 8 = \dfrac{-1}{8}(x - 5)$

EXAMPLE 10: The curve C is defined by the parametric equations $x = e^{-t} + 1$ and $y = \sin \pi t - t^2$.

a) At $t = 2$, is the curve increasing, decreasing or neither? Justify your answer.

b) At $t = 2$, is the curve concave up, concave down or neither? Justify your answer.

SOLUTIONS:

a) $\dfrac{dy}{dx} = \dfrac{\dfrac{dy}{dt}}{\dfrac{dx}{dt}} = \dfrac{\pi \cos \pi t - 2t}{-e^{-t}} = \dfrac{\pi \cos \pi t - 2t}{\dfrac{-1}{e^t}} = -e^t \left(\pi \cos \pi t - 2t \right)$

$\dfrac{dy}{dx}_{[t=2]} = -e^2 (\pi - 4) > 0$ so C is increasing.

b) $\dfrac{d^2 y}{dx^2} = \dfrac{\dfrac{d}{dt}\left(\dfrac{dy}{dx}\right)}{\dfrac{dx}{dt}} = \dfrac{-e^t \left(-\pi^2 \sin \pi t - 2 + \pi \cos \pi t - 2t \right)}{-e^{-t}}$

$= e^{2t} \left(-\pi^2 \sin \pi t - 2 + \pi \cos \pi t - 2t \right)$

$\dfrac{d^2 y}{dx^2}_{[t=2]} = e^4 (-2 + \pi - 4) < 0$ so C is concave down.

TEST TIP

Your calculator can graph parametric equations. Change your MODE to PAR, input your equations and adjust your window. You need to set the minimum and maximum values for t as well as the Tstep.

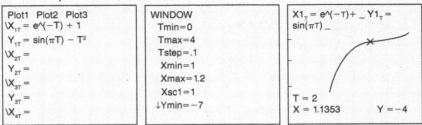

Note in Example 10 that if you trace the curve C, it will graph from right to left, which might make you think that the graph is decreasing. But, remember, this is not a particle motion problem. You are being asked about the shape of the graph and the behavior at $t = 2$. It is increasing and concave down. A particle moving along this curve would be moving left as $\dfrac{dx}{dt} < 0$, but the curve itself is increasing because $\dfrac{dy}{dx} > 0$.

EXAMPLE 11: (Calculator Active) A curve C has the property that at every point (x,y) on C, $x(t) = \sqrt{t}$ and $y(t)$ is not specifically given. If $\dfrac{dy}{dt} = \ln t$ and $\lim\limits_{t \to 0^+} y(t) = -4$,

a) Find the point on C that corresponds to $t = 2$.

b) Find the length of the arc of curve C from $t = 1$ to $t = 2$.

SOLUTIONS:

$$y(t) = \int \ln t\, dt$$

$$y(t) = t \ln t - \int \left(\frac{t}{t}\right) + C \qquad u = \ln t \qquad v = t$$

$$y(t) = t \ln t - t + C \qquad du = \frac{1}{t} dt \quad dv = dt$$

$$\lim_{t \to 0} y(t) = 0(-\infty) - 0 + C$$

Note that $0\,(-\infty)$ is an indeterminate form, so L'Hospital's Rule must be used.

a) $\lim\limits_{t \to 0} \dfrac{\ln t}{\dfrac{1}{t}} - t + C = \lim\limits_{t \to 0} \dfrac{\dfrac{1}{t}}{\dfrac{-1}{t^2}} - t + C$

$\lim\limits_{t \to 0} (-t - t + C) = -4 \Rightarrow C = -4$

$y(t) = t \ln t - t - 4$

$y(2) = 2 \ln 2 - 2 - 4 = 2 \ln 2 - 6 \text{ or } \ln 4 - 6$

$x(2) = \sqrt{2}$

Point: $\left(\sqrt{2},\, 2\ln 2 - 6\right) \text{ or } \left(\sqrt{2},\, \ln 4 - 6\right)$

b) $\text{Dist} = \int\limits_{1}^{2} \sqrt{\left(\dfrac{1}{2\sqrt{t}}\right)^2 + (\ln t)^2}\, dt$

$\text{Dist} \approx 0.594$

TEST TIP

Example 11 cleverly incorporates integration by parts and L'Hospital's rule to a parametric problem. Even if you cannot do part a, try doing part b as it is independent of part a. This type of situation occurs many times on the free-response section of the exam, so students should read and attempt every part of a problem.

Vector-Valued Functions

Overview: Quantities such as area, volume, temperature, and time can be characterized by a single real number called a **scalar**. Quantities such as force, velocity and direction cannot be represented by scalars; they need to be represented by vectors. Vectors are **directed line segments** involving both magnitude and direction. While courses in multivariable calculus do an extensive job of teaching vectors, in BC calculus, vectors are little more than parametric equations in disguise.

Typically, instead of an object moving along a straight line, you will be given a situation where an object is moving in the plane. Its position in the plane can be written as a vector: $\mathbf{r}$ where the position vector: $\mathbf{r}(t) = \langle x(t), y(t) \rangle$. From there, you can find its velocity vector: $\mathbf{v}(t) = \mathbf{r}'(t) = \langle x'(t), y'(t) \rangle$ and its acceleration vector: $\mathbf{a}(t) = \mathbf{v}'(t) = \mathbf{r}''(t) = \langle x''(t), y''(t) \rangle$.

Or, we can use integration starting with the acceleration vector and find the velocity vector: $\mathbf{v}(t) = \int \mathbf{a}(t)\, dt + \mathbf{C}\langle C_x, C_y \rangle$. $\mathbf{C}\langle C_x, C_y \rangle$ refers to a constant vector. And we can find the position vector by integrating: position vector: $\mathbf{r}(t) = \int \mathbf{v}(t)dt + \mathbf{C}\langle C_x, C_y \rangle$.

The **magnitude** refers to the length of a vector. The magnitude of vector $\mathbf{v}\langle x(t), y(t) \rangle$, written $|\mathbf{v}|$, is $\sqrt{[x(t)]^2 + [y(t)]^2}$. Students should also know that the **speed** of the object is defined as the absolute value of the velocity: $|\mathbf{v}(t)| = \sqrt{[x'(t)]^2 + [y'(t)]^2}$. The magnitude and speed are scalars, not vectors.

EXAMPLE 12: An object moving along a curve in the xy-plane has position $(x(t), y(t))$ at time t with

$$\frac{dx}{dt} = 4\sin 4t \quad \text{and} \quad \frac{dy}{dt} = 6t - 3 \quad \text{for } t \geq 0.$$

a) What is the magnitude of the object's acceleration at $t = \pi$?

b) At time $t = 0$, the object is at position $(1, \pi)$. Where is the object at $t = \pi$?

c) For how many values of t, $0 \leq t \leq \pi$, is the tangent to the curve vertical?

SOLUTIONS: a) $\mathbf{a}(t) = \langle 16\cos 4t, 6 \rangle \Rightarrow \mathbf{a}(\pi) = \langle 16, 6 \rangle$

$$|\mathbf{a}(\pi)| = \sqrt{16^2 + 6^2} = \sqrt{256 + 36} = \sqrt{292} = 2\sqrt{73}$$

b)

$x = -\cos 4t + C_1 \Rightarrow x(0) = -1 + C_1 = 1 \Rightarrow C_1 = 2 \Rightarrow x = -\cos 4t + 2$

$y = 3t^2 - 3t + C_2 \Rightarrow y(0) = C_2 = \pi \Rightarrow y = 3t^2 - 3t + \pi$

$x(\pi) = -1 + 2 = 1 \qquad y(\pi) = 3\pi^2 - 3\pi + \pi = 3\pi^2 - 2\pi$

$(x, y)_{t=\pi} = (1, 3\pi^2 - 2\pi)$

c) The tangent line is vertical when $\dfrac{dx}{dt} = 0$ and $\dfrac{dy}{dt} \neq 0$.

This occurs when $4\sin 4t = 0$ or $t = 0, \dfrac{\pi}{4}, \dfrac{\pi}{2}, \dfrac{3\pi}{4}, \pi$ which are 5 values.

EXAMPLE 13: The position of a particle moving in the xy-plane with position function $\mathbf{r}(t) = \langle e^{-t} - \sin t, \cos t \rangle$, when $t \geq 0$. What is the maximum speed attained by the particle?

SOLUTION: $x'(t) = -e^{-t} - \cos t \qquad y'(t) = -\sin t$

$$|\mathbf{v}(t)| = \sqrt{[x'(t)]^2 + [y'(t)]^2} = \sqrt{e^{-2t} + 2e^{-t}\cos t + \cos^2 t + (-\sin t)^2}$$

$$|\mathbf{v}(t)| = \sqrt{e^{-2t} + 2e^{-t}\cos t + \cos^2 t + \sin^2 t}$$

Note: $\cos^2 t + \sin^2 t = 1$ for all values of t.

The maximum value of $e^{-2t} = 1$ occurs at $t = 0, 2\pi, 4\pi \ldots$,

so the maximum value of $|\mathbf{v}(t)| = \sqrt{1 + 2\cos t + 1} = \sqrt{2 + 2\cos t}$

At $t = 0, 2\pi, 4\pi \ldots, \sqrt{2 + 2\cos t} = 2$.

EXAMPLE 14: The figure on the next page models the path of a raindrop moving along a windshield for the time interval, $0 \leq t \leq 2\pi$, t measured in seconds. The path of the raindrop is modeled by the position vector: $\mathbf{h}(t) = \left\langle (8 - t)\sin t, 9 - \dfrac{t^2}{2} \right\rangle$, measured in inches.

a) Find the slope of the path of the raindrop at $t = \pi$.

b) Find the speed of the raindrop in inches/sec at $t = \pi$.

c) Write an expression to find the average speed for the raindrop over the interval $0 \leq t \leq 2\pi$.

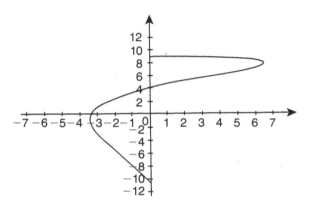

SOLUTIONS: a) $\mathbf{v}(t) = \langle (8-t)\cos t - \sin t, -t \rangle$

$$\frac{dy}{dx} = \frac{dy/dt}{dx/dt}\bigg|_{t=\pi} = \frac{-\pi}{\pi-8} = \frac{\pi}{8-\pi}$$

b) Speed $= \sqrt{[x'(t)]^2 + [y'(t)]^2}$

Speed$_{t=\pi} = \sqrt{(\pi-8)^2 + (-\pi)^2} = \sqrt{2\pi^2 - 16\pi + 64}$ inches/sec.

c) Avg speed $= \dfrac{1}{2\pi} \displaystyle\int_0^{2\pi} \sqrt{[x'(t)]^2 + [y'(t)]^2}\, dt$

$$\frac{1}{2\pi} \int_0^{2\pi} \sqrt{((8-t)\cos t - \sin t)^2 + (-t)^2}\, dt \text{ inches/sec.}$$

Polar Equations

Overview: Points using **polar coordinates** are a given radius r from the pole (origin) and θ, the central angle at the pole. A polar point is in the form (r, θ) where r is the radius, the distance from the pole and θ is the angle (in radians) from the x-axis in a counterclockwise direction. The points A $(3,0)$, $B\left(2, \dfrac{\pi}{3}\right)$, $C\left(4, -\dfrac{5\pi}{6}\right)$, and $D\left(\dfrac{-5}{2}, \dfrac{\pi}{2}\right)$ are shown in the figure below. Note that there is more than one way to name a point using polar coordinates. For example, point D can also be named as $\left(\dfrac{5}{2}, \dfrac{3\pi}{2}\right)$. Likewise, point A can also be named as $(3, 2\pi)$.

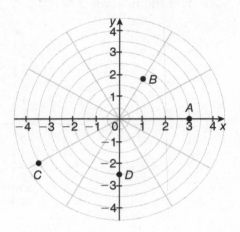

Polar equations are in the form $r = f(\theta)$, a set of radii given as a function of θ. Polar graphs can be quite complicated and quite beautiful as they are not necessarily functions in the normal sense of the vertical-line test.

We usually are asked to find slopes of these curves, which means that we need to represent the polar curve parametrically. The formulas to do so are: $x = r\cos\theta$ and $y = r\sin\theta$. So the polar equation $r = 1 + \sin\theta$ can be expressed parametrically as:

$$x = (1 + \sin t)\cos t \text{ and } y = (1 + \sin t)\sin t.$$

To find the slope $\dfrac{dy}{dx}$ of an equation expressed in polar form, you use the formula: $\dfrac{dy}{dx} = \dfrac{dy/d\theta}{dx/d\theta}$. Horizontal tangents occur when $\dfrac{dy}{d\theta} = 0$ and vertical tangents occur when $\dfrac{dx}{d\theta} = 0$. If $\dfrac{dy}{d\theta}$ and $\dfrac{dx}{d\theta}$ are both zero simultaneously, no conclusion can be made.

Typical AP problems involve finding the area bounded by a polar curve between two angles α and β.

$A = \dfrac{1}{2}\displaystyle\int_{\alpha}^{\beta}[f(\theta)]^2 d\theta$. This assumes the function f is continuous and non-negative.

Arc length in polar form is given by the formula:

$$s = \int_{\alpha}^{\beta}\sqrt{[f(\theta)]^2 + [f'(\theta)]^2}\,d\theta = \int_{\alpha}^{\beta}\sqrt{r^2 + \left(\dfrac{dr}{d\theta}\right)^2}\,d\theta.$$

EXAMPLE 15: (Calculator Active) The figure below shows the graph for $r = 1 + \cos 2\theta$ for $0 \le \theta \le 2\pi$. Find the area enclosed by the graph.

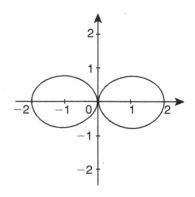

SOLUTION: $A = \dfrac{1}{2} \displaystyle\int_{0}^{2\pi} r^2 d\theta = \dfrac{1}{2} \int_{0}^{2\pi} [1 + \cos(2\theta)]^2 d\theta \approx 4.712$

$A = \displaystyle\int_{0}^{\pi} r^2 d\theta$ or $A = 2 \displaystyle\int_{0}^{\pi/2} r^2 d\theta$ is also acceptable.

EXAMPLE 16: What is the equation of the line tangent to the polar curve $r = 2\sin\theta + \cos\theta$ at $\theta = \dfrac{3\pi}{4}$?

SOLUTION:

You need to expressed the polar equation parametrically.

$x = r\cos\theta = 2\sin\theta\cos\theta + \cos^2\theta$ $\qquad\qquad$ $y = r\sin\theta = 2\sin^2\theta + \sin\theta\cos\theta$

$x\left(\dfrac{3\pi}{4}\right) = 2\left(\dfrac{\sqrt{2}}{2}\right)\left(\dfrac{-\sqrt{2}}{2}\right) + \left(\dfrac{-\sqrt{2}}{2}\right)^2 = \dfrac{-1}{2}$ $\qquad$ $y\left(\dfrac{3\pi}{4}\right) = 2\left(\dfrac{\sqrt{2}}{2}\right)^2 + \left(\dfrac{\sqrt{2}}{2}\right)\left(\dfrac{-\sqrt{2}}{2}\right) = \dfrac{1}{2}$

$\dfrac{dx}{d\theta} = 2(-\sin^2\theta + \cos^2\theta) - 2\cos\theta\sin\theta$ $\qquad$ $\dfrac{dy}{d\theta} = 4(\sin\theta\cos\theta) + (-\sin^2\theta + \cos^2\theta)$

$\dfrac{dy}{dx}\Big|_{\theta=3\pi/4} = \dfrac{\dfrac{dy}{d\theta}}{\dfrac{dx}{d\theta}} = \dfrac{4\left(\dfrac{-1}{2}\right) + \left(-\dfrac{1}{2} + \dfrac{1}{2}\right)}{2\left(-\dfrac{1}{2} + \dfrac{1}{2}\right) - 2\left(\dfrac{-\sqrt{2}}{2}\right)\left(\dfrac{\sqrt{2}}{2}\right)} = \dfrac{-2}{1} = -2$

Tangent Line: $y - \dfrac{1}{2} = -2\left(x + \dfrac{1}{2}\right)$ or $y = -2x - \dfrac{1}{2}$

EXAMPLE 17:   A curve shown in the figure below is described by the polar equation $r = 1 + \theta + 2\sin\theta$.

a) Find $\dfrac{dr}{d\theta}$ at $\theta = \dfrac{5\pi}{6}$. Interpret the meaning of this value with respect to the curve below.

b) Find the value of θ where $0 \le \theta \le \pi$ that corresponds to the point on the curve with greatest distance from the origin. Justify your answer.

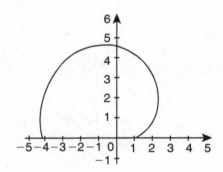

SOLUTIONS:

a) $\dfrac{dr}{d\theta} = 1 + 2\cos(\theta) \Rightarrow \dfrac{dr}{d\theta}\Big|_{\theta = \frac{5\pi}{6}} = 1 + 2\cos\dfrac{5\pi}{6} = 1 + 2\left(\dfrac{-\sqrt{3}}{2}\right) = 1 - \sqrt{3}$

At $\theta = \dfrac{5\pi}{6}$, r is decreasing meaning the curve is getting closer to the origin.

b) $\dfrac{dr}{d\theta} = 1 + 2\cos(\theta) = 0 \Rightarrow \cos(\theta) = \dfrac{-1}{2} \Rightarrow \theta = \dfrac{2\pi}{3}$ or $\dfrac{4\pi}{3}$.

Ignore the value of $\dfrac{4\pi}{3}$ because it lies outside $0 \le \theta \le \pi$.

θ	0	$\dfrac{2\pi}{3}$	π
r	1	$1 + \dfrac{2\pi}{3} + \sqrt{3}$	$1 + \pi$

The greatest distance occurs when $\theta = \dfrac{2\pi}{3}$.

TEST TIP

When calculators are permitted on the AP exam, graphs like this can be created by setting the MODE to Polar, inputting the equation, and setting an appropriate window. Using 2nd FORMAT, you should change to Polar graphing coordinates. All graph traces or information from 2nd CALC will show information using R and θ. Of course, θ will not be in terms of π.

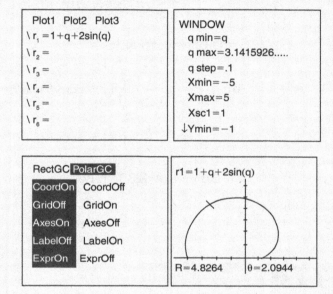

EXAMPLE 18: (Calculator Active) A road race takes place on a course whose path is the graph of $r = 1 + 2\cos\theta$ shown below where r is measured in miles. The course has two loops. The adult race is the outer loop and the children's race is the inner loop around a lake.

a) Find the difference between the length of the adult race and the length of the children's race.

b) Find the area between the adult's course and the lake.

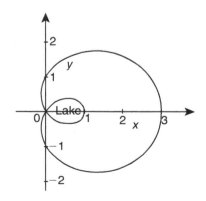

SOLUTIONS: $1 + 2\cos\theta = 0 \Rightarrow \cos\theta = -\dfrac{1}{2} \Rightarrow \theta = \dfrac{2\pi}{3}, \dfrac{4\pi}{3}$

a) $s_{\text{adult}} = 2\displaystyle\int_{0}^{2\pi/3} \sqrt{(1+2\cos\theta)^2 + (-2\sin\theta)^2}\ d\theta \approx 10.682$

$s_{\text{child}} = 2\displaystyle\int_{2\pi/3}^{\pi} \sqrt{(1+2\cos\theta)^2 + (-2\sin\theta)^2}\ d\theta \approx 2.682$

Difference: 8 miles

b) $A_{\text{outer}} = 2\left(\dfrac{1}{2}\right)\displaystyle\int_{0}^{2\pi/3} (1+2\cos\theta)^2\ d\theta \approx 8.881$

$A_{\text{inner}} = 2\left(\dfrac{1}{2}\right)\displaystyle\int_{2\pi/3}^{\pi} (1+2\cos\theta)^2\ d\theta \approx .544$

Difference: 8.337 miles^2

TEST TIP

If students were only asked to set up integrals for Example 18, no calculator would be needed. However, if a problem of this nature is on the AP exam and a calculator is allowed, you might want to graph it in degree mode. When you TRACE in order to find values where $r = 0$, the angle will be in degrees which are easily changed to radians. However, when you compute the value of the integral, be sure to change back to radian mode or your answer will be incorrect.

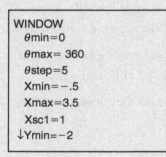

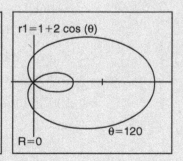

EXAMPLE 19: (Calculator Active) The graph of $r = 6\cos\theta$ is a circle as shown below. Find the area bounded by the graph in three different ways.

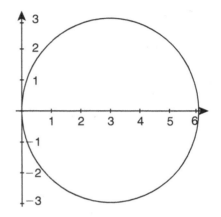

SOLUTION: Area of a circle: $\pi r^2 = 9\pi \approx 28.274$

Polar coordinates: $A = \dfrac{1}{2}\displaystyle\int_0^\pi 36\cos^2\theta\, d\theta \approx 28.274$

Rectangular Coordinates: $r = 6\cos\theta \Rightarrow r = 6\left(\dfrac{x}{r}\right)$

$r^2 = 6x \Rightarrow x^2 + y^2 = 6x \Rightarrow y = \sqrt{6x - x^2}$

$A = 2\displaystyle\int_0^6 \sqrt{6x - x^2}\, dx \approx 28.274$

```
.5 fnInt (36(cos(X
))²,X,θ,π)
            28.27433388
2fnInt(√(6X−X²),
X,0,6)
            28.27433479
```

DIDYOUKNOW?

Take a pizza and pick any arbitrary point within it. Cut the pizza into 8 slices by cutting at 45-degree angles through that point and label the alternate pieces A and B. Person A gets the A slices and person B gets the B slices. Surprising result: A and B always get half of the pizza. This can be proven using polar coordinates and calculus.

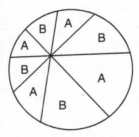

Time for a quiz
• Review strategies in Chapter 2
• Take Quiz 8 at the REA Study Center
(www.rea.com/studycenter)

Sequences and Series

Overview: A **sequence** is a set of numbers that has an identified first member, second member, third member, etc. We typically use subscript notation to define the terms: $a_1, a_2, a_3, \ldots, a_n$.

The sequence can be given three different ways:

- The general term a_n can be given. The student is expected to write the first few terms of the sequence. For example, if $a_n = \dfrac{n}{n+1}$, the sequence is $\left\{ \dfrac{1}{2}, \dfrac{2}{3}, \dfrac{3}{4}, \dfrac{4}{5}, \ldots \right\}$.

- The first few terms of the sequence can be given and it is up to the student to generate further terms and a formula for the general term a_n.

  For instance, a sequence might be given as $\left\{ \dfrac{1}{2}, \dfrac{1}{4}, \dfrac{1}{8}, \ldots \right\}$. The general term appears to be $a_n = \dfrac{1}{2^n}$, but it is not necessarily clear. Believe it or not, the general term could be

  $a_n = \dfrac{6}{(n+1)(n^2 - n + 6)}$.

- The sequence could be defined **recursively** where the $(n+1)$st term is dependent on the nth term. For instance, if $a_1 = 3$ and $a_{n+1} = 2\,a_n - 1$, the sequence would be $\{3, 5, 9, 17, \ldots\}$. AP exam questions rarely use this form.

A sequence is said to **converge** if its terms approach a limiting value. Otherwise, the sequence is said to **diverge**. For example the sequence $\left\{\dfrac{1}{2},\dfrac{2}{3},\dfrac{3}{4},\dfrac{4}{5},\ldots\dfrac{n}{n+1}\right\}$ converges because $\lim\limits_{n\to\infty}\dfrac{n}{n+1}=1$ (by L'Hospital's Rule). The sequence $\{-2,4,-8,16,\ldots(-2)^n\}$ diverges because the terms alternate and its terms do not approach a limiting value.

A **series** is the sum of a sequence. Calculus explores **infinite series**: the sum of an infinite number of terms. We usually write this using what is called Sigma Notation (using the Greek letter Sigma Σ): $\displaystyle\sum_{n=1}^{\infty} a_n = a_1 + a_2 + a_3 + \ldots + a_n + \ldots$

We are interested in **convergent series**, that is: $\displaystyle\sum_{n=1}^{\infty} a_n = a_1 + a_2 + a_3 + \ldots + a_n + \ldots = L$ where L is the limit of the sum of the infinite terms.

Important: A convergent sequence has its nth-term approaching a limit while a convergent series has the *sum of its infinite terms* approaching a limit.

If $\lim\limits_{n\to\infty} a_n \neq 0$, it is impossible for the series to converge. For example, $\displaystyle\sum_{n=1}^{\infty}\dfrac{n}{n+1} = \dfrac{1}{2} + \dfrac{2}{3} + \dfrac{3}{4} + \ldots$ could not converge because $\lim\limits_{n\to\infty}\dfrac{n}{n+1}=1$. As we add more and more terms, we are essentially adding a larger amount each time and the series gets infinitely large.

However, it is important to understand that if $\lim\limits_{n\to\infty} a_n = 0$, that does not necessarily mean that a series converges.

This is summarized by what is known as the ***n*th-term test**:

If $\lim\limits_{n\to\infty} a_n \neq 0$, the series a_n *cannot* converge.

If $\lim\limits_{n\to\infty} a_n = 0$, the series a_n *might* converge. More investigation is necessary.

EXAMPLE 1:   Consider the sequence $\{a_n\} = \left\{\dfrac{n}{1-4n}\right\}$.

   a) Determine whether the sequence converges or diverges.

   b) Determine whether the series for $\{a_n\}$ converges or diverges.

SOLUTIONS: a) $\{a_n\} = \left\{\dfrac{n}{1-4n}\right\} = \left\{\dfrac{-1}{3}, \dfrac{-2}{7}, \dfrac{-3}{11}...\right\}$. Since $\lim\limits_{n\to\infty} a_n = \dfrac{-1}{4}$ (by

   L'Hospital's Rule), the sequence converges.

   b) $\sum\limits_{n=1}^{\infty} a_n = -\dfrac{1}{3} - \dfrac{2}{7} - \dfrac{3}{11} - ....$ Since $\lim\limits_{n\to\infty} a_n = \dfrac{-1}{4} \neq 0$, by the

   nth-term test, the series diverges.

EXAMPLE 2:   Consider the sequence $\{b_n\} = \{(-1)^{n+1}\}$.

   a) Determine whether the sequence converges or diverges.

   b) Determine whether the series for $\{b_n\}$ converges or diverges.

SOLUTIONS: a) $\{b_n\} = \{(-1)^{n+1}\} = \{1, -1, 1, -1,...\}$. Since $\lim\limits_{n\to\infty} b_n$ does not exist,
   the sequence diverges.

   b) $\sum\limits_{n=1}^{\infty} b_n = 1 - 1 + 1 - 1 + ...$ Since $\lim\limits_{n\to\infty} b_n \neq 0$, the series also
   diverges.

DIDYOU**KNOW?**

Students typically think that the series in Example 2b converges. The thought is that for $1-1+1-1+...$, every two terms cancel out and $\sum_{n=1}^{\infty} b_n = 0$. However, with that thinking, we can say that $1-1+1-1+... = 1+(-1+1)+(-1+1) + ...= 1$. So if $\sum_{n=1}^{\infty} b_n = 0$, then $\sum_{n=1}^{\infty} b_n = 1$, implying 0 = 1, which is clearly impossible. That is why $\sum_{n=1}^{\infty} b_n$ diverges.

EXAMPLE 3: Consider the sequence $\{c_n\} = \left\{\dfrac{n^2}{10^n - n + 1}\right\}$.

a) Determine whether the sequence converges or diverges.

b) Determine whether the series for $\{c_n\}$ converges or diverges.

SOLUTIONS: a) $\{c_n\} = \left\{\dfrac{1}{10}, \dfrac{4}{99}, \dfrac{9}{998}, ...\right\}$. Since by L'Hospital's Rule,

$$\lim_{n\to\infty} \frac{n^2}{10^n - n + 1} = \lim_{n\to\infty} \frac{2n}{10^n \ln 10 - 1} = \lim_{n\to\infty} \frac{2}{10^n (\ln 10)^2} = \lim_{n\to\infty} \frac{0}{10^n (\ln 10)^3} = 0,$$

the sequence converges.

b) Since $\lim_{n\to\infty} \dfrac{n^2}{10^n - n + 1} = 0$, the series could converge. At this point, we do not know and more testing is necessary.

Basic Series and Convergence Tests

There are a number of general series that students need to recognize, whether or not they converge, and if so, what values make it converge:

Telescoping Series: The telescoping series is in the form

$$\sum_{n=1}^{\infty} (a_{n+1} - a_n) = a_2 - a_1 + a_3 - a_2 + a_4 - a_3 + ... \text{ which is always convergent. The series}$$

$$\sum_{n=1}^{\infty}\frac{1}{n^2+n}=\sum_{n=1}^{\infty}\left(\frac{1}{n}-\frac{1}{n+1}\right)=1-\frac{1}{2}+\frac{1}{2}-\frac{1}{3}+\frac{1}{3}-\frac{1}{4}+...\text{ clearly (with a little help}$$

from partial fraction decomposition) converges to 1 because all terms except the first cancel.

Geometric Series: A geometric series is in the form of

$$\sum_{n=0}^{\infty}ar^n=a+ar+ar^2+...+ar^n+.... \text{ A geometric series will diverge if }|r|\geq 1\text{ and will}$$

converge if $|r|<1$. The geometric and telescoping series are the only series that students

must know what the series will converge to. For geometric series: $\sum_{n=0}^{\infty}ar^n=\dfrac{a}{1-r}$.

- The geometric series $\sum_{n=0}^{\infty}\dfrac{5}{3^n}=\sum_{n=0}^{\infty}5\left(\dfrac{1}{3}\right)^n=5+\dfrac{5}{3}+\dfrac{5}{9}...$ converges as $r=\dfrac{1}{3}<1$. It

  converges to $\dfrac{5}{1-\dfrac{1}{3}}=\dfrac{15}{2}$.

- The geometric series $\sum_{n=0}^{\infty}\left(\dfrac{10}{9}\right)^n=1+\dfrac{10}{9}+\dfrac{100}{81}+...$ diverges as $\dfrac{10}{9}>1$.

- The series $8-6+\dfrac{9}{2}-\dfrac{27}{8}+...$ can be written as

  $8\left(1-\dfrac{6}{8}+\dfrac{9}{16}-\dfrac{27}{64}+...\right)=8\sum_{n=0}^{\infty}\left(\dfrac{-3}{4}\right)^n$. This is a geometric series with

  $r=\dfrac{-3}{4}$ and $|r|<1$. The sum of the series is $\dfrac{8}{1-\left(\dfrac{-3}{4}\right)}=\dfrac{32}{7}$.

Harmonic Series: A harmonic series is in the form $\sum_{n=1}^{\infty}\dfrac{1}{an+b}$ where a and b are

constants. Harmonic series always diverge. The simplest harmonic series that always

shows up on an AP exam is $\sum_{n=1}^{\infty}\dfrac{1}{n}=1+\dfrac{1}{2}+\dfrac{1}{3}+...$ which is divergent.

***p*-series:** A p-series is in the form $\sum_{n=1}^{\infty}\dfrac{1}{n^p}=1+\dfrac{1}{2^p}+\dfrac{1}{3^p}+...$ This series will converge

when $p>1$ and diverges when $0<p\leq 1$. When $p=1$, we get the divergent harmonic

series $\sum_{n=1}^{\infty}\dfrac{1}{n}$.

- $\sum_{n=1}^{\infty} \frac{1}{n^2} = 1 + \frac{1}{2^2} + \frac{1}{3^2} + \ldots = 1 + \frac{1}{4} + \frac{1}{9} + \ldots$ is convergent because $p > 1$.

- $\sum_{n=1}^{\infty} \frac{1}{n^{1/2}} = 1 + \frac{1}{\sqrt{2}} + \frac{1}{\sqrt{3}} + \ldots$ is divergent because $p \leq 1$.

Students can confuse a p-series with a geometric series. In a geometric series, a constant term is being raised to successively higher powers while in a p-series, the terms change, but the power remains the same.

Alternating Series: An alternating series is in the form of

$\sum_{n=1}^{\infty} (-1)^n a_n$ or $\sum_{n=1}^{\infty} (-1)^{n+1} a_n$. The terms alternate in sign. They converge if $\lim_{n \to \infty} a_n = 0$

(nth-term test) and $a_{n+1} < a_n$ for all n. This means that the terms themselves (without regard to sign) are getting successively smaller.

- $\sum_{n=1}^{\infty} (-1)^{n+1} \frac{1}{n} = 1 - \frac{1}{2} + \frac{1}{3} - \frac{1}{4} + \frac{1}{5} \ldots$ is the alternating harmonic series and it is convergent although the harmonic series $\sum_{n=1}^{\infty} \frac{1}{n}$ is divergent.

- $\sum_{n=1}^{\infty} (-1)^{n+1} \frac{n}{n+1} = \frac{1}{2} - \frac{2}{3} + \frac{3}{4} - \frac{4}{5} + \ldots$ is divergent because it fails the nth-term test $\left(\lim_{n \to \infty} \frac{n}{n+1} = 1 \neq 0 \right)$.

- $\sum_{n=1}^{\infty} \frac{\cos \pi n}{n^2}$ is an alternating series in disguise. $\sum_{n=1}^{\infty} \frac{\cos \pi n}{n^2} = -1 + \frac{1}{4} - \frac{1}{9} + \frac{1}{16} - \ldots$ It is convergent.

When approximating the sum of a converging alternating series using n terms, there is an error associated with that approximation. This error is simply the $|(n+1)^{st}|$ term: $|a_{n+1}|$.

- The error in approximating the convergent alternating series $\sum_{n=1}^{\infty} (-1)^{n+1} \frac{1}{n!}$ using 5 terms is simply $\frac{1}{6!} = \frac{1}{720}$.

DIDYOUKNOW?

Did you know that when adding numbers, it is possible that the order you add them may matter? For instance, start with the alternating harmonic series: $1 - \frac{1}{2} + \frac{1}{3} - \frac{1}{4} + \ldots$ We know this is convergent (and it converges to ln 2). However, it can be shown that this series can be said to converge to any number you want. For instance, let's show it is convergent to 4. Take enough positive terms and add them (in their original order) so that they add to just above 4. Now add on enough negative terms so that the partial sum dips below 4. Now add more positive terms so that the sum is now above 4 and continue with this process infinitely. This arrangement of adding will sum to 4. As mentioned previously in the text, when working with an infinite number of numbers, strange things can happen and logic goes out the window.

When series resemble but do not precisely match the definitions of the series above, there are convergence tests that are used that will usually determine whether or not a series is convergent. The Limit Comparison Test and Ratio Test handle just about all situations but, occasionally, both fail, and the Integral Test will usually suffice.

The Limit Comparison Test: If two series $a_n, b_n > 0$ and $\lim\limits_{n \to \infty} \left(\frac{a_n}{b_n} \right) = L$ where L is both finite and positive, then $\Sigma\, a_n$ and $\Sigma\, b_n$ either both converge or both diverge. What this says is that when you are interested in whether a series a_n converges, choose a series b_n which you know is either convergent or divergent and find $\lim\limits_{n \to \infty} \left(\frac{a_n}{b_n} \right)$. If you get a positive finite answer, the a_n series does exactly what the b_n series does.

- $\sum\limits_{n=1}^{\infty} \frac{1}{n^2 + 4n - 2}$ is convergent as we compare it to the convergent p-series $\sum\limits_{n=1}^{\infty} \frac{1}{n^2}$.

$\lim\limits_{n \to \infty} \dfrac{\frac{1}{n^2 + 4n - 2}}{\frac{1}{n^2}} = \lim\limits_{n \to \infty} \dfrac{n^2}{n^2 + 4n - 2} = 1$ which is positive and finite. Note that we are not saying that $\sum\limits_{n=1}^{\infty} \frac{1}{n^2 + 4n - 2}$ is convergent to 1, just that it is convergent.

- $\displaystyle\sum_{n=1}^{\infty}\frac{2n^2+3}{n^3-n+5}$ is divergent as we compare it to the divergent harmonic series $\displaystyle\sum_{n=1}^{\infty}\frac{1}{n}$.

$$\lim_{n\to\infty}\frac{\dfrac{2n^2+3}{n^3-n+5}}{\dfrac{1}{n}}=\lim_{n\to\infty}\frac{2n^3+3n}{n^3-n+5}=2 \text{ which is positive and finite. Since }\sum_{n=1}^{\infty}\frac{1}{n}\text{ diverges,}$$

so does $\displaystyle\sum_{n=1}^{\infty}\frac{2n^2+3}{n^3-n+5}$.

The Ratio Test: This is the most widely used convergence test on the AP exam.

$\displaystyle\sum a_n$ converges if $\displaystyle\lim_{n\to\infty}\left|\frac{a_{n+1}}{a_n}\right|<1$.

$\displaystyle\sum a_n$ diverges if $\displaystyle\lim_{n\to\infty}\left|\frac{a_{n+1}}{a_n}\right|>1$ or $\displaystyle\lim_{n\to\infty}\left|\frac{a_{n+1}}{a_n}\right|=\infty$

The ratio test is inconclusive if $\displaystyle\lim_{n\to\infty}\left|\frac{a_{n+1}}{a_n}\right|=1$.

- The series $\displaystyle\sum_{n=0}^{\infty}\frac{4^n}{n!}=1+4+\frac{16}{2}+\frac{64}{6}+...$ appears divergent as the terms are not getting smaller. However, convince yourself that this series could converge because $n!$ is ultimately more powerful than 4^n. The series passes the nth-term test. But that doesn't mean the series converges. Using the ratio test will confirm that it does.

$$\lim_{n\to\infty}\left|\frac{a_{n+1}}{a_n}\right|=\lim_{n\to\infty}\left|\frac{\dfrac{4^{n+1}}{(n+1)!}}{\dfrac{4^n}{n!}}\right|=\lim_{n\to\infty}\left|\frac{4^{n+1}}{(n+1)!}\times\frac{n!}{4^n}\right|=\lim_{n\to\infty}\left|\frac{4}{n+1}\right|=0<1.$$

- The series $\displaystyle\sum_{n=1}^{\infty}n\left(\frac{4}{5}\right)^n=\frac{4}{5}+2\left(\frac{4}{5}\right)^2+3\left(\frac{4}{5}\right)^3...=\frac{4}{5}+\frac{16}{25}+\frac{192}{125}+...$ may appear divergent because beginning with the third term, the numerator is greater than the denominator. But looks are deceiving. By the ratio test:

$$\lim_{n\to\infty}\left|\frac{a_{n+1}}{a_n}\right|=\lim_{n\to\infty}\left|\frac{(n+1)\left(\dfrac{4}{5}\right)^{n+1}}{n\left(\dfrac{4}{5}\right)^n}\right|=\lim_{n\to\infty}\left|\frac{n+1}{n}\times\left(\frac{4}{5}\right)\right|=\frac{4}{5}<1 \text{ so the series }\sum_{n=1}^{\infty}n\left(\frac{4}{5}\right)^n$$

is convergent.

- The series $\sum_{n=1}^{\infty}(-1)^{n+1}\dfrac{2^n}{n^5}=2-\dfrac{4}{32}+\dfrac{8}{343}-\dfrac{16}{1024}+...$ appears as if it could converge because the denominator seems to be greater than the numerator. But again, looks are deceiving and this series fails the nth-term test. (Plug in $n=30$ to convince yourself.) But the ratio test cuts through the trial and error.

$$\lim_{n\to\infty}\left|\frac{a_{n+1}}{a_n}\right|=\lim_{n\to\infty}\left|\frac{(-1)^{n+2}\dfrac{2^{n+1}}{(n+1)^5}}{(-1)^{n+1}\dfrac{2^n}{n^5}}\right|=\lim_{n\to\infty}\left|(-1)\times\frac{2^{n+1}}{(n+1)^5}\times\frac{n^5}{2^n}\right|=\lim_{n\to\infty}\left|(-1)\times\frac{n^5}{(n+1)^5}\times\frac{2^{n+1}}{2^n}\right|=2>1$$

so the series $\sum_{n=1}^{\infty}(-1)^{n+1}\dfrac{2^n}{n^5}$ diverges.

The Integral Test: If f is positive, continuous and decreasing and $a_n=f(n)$, then $\sum_{n=1}^{\infty}a_n$ and $\int_{1}^{\infty}f(x)\,dx$ either both converge or diverge.

- Proving divergence for the series $\sum_{n=2}^{\infty}\dfrac{\ln n}{n}$ does not work with the limit comparison test as

$\lim_{n\to\infty}\left(\dfrac{\ln n}{n}\times\dfrac{n}{1}\right)=\infty$ and does not work with the ratio test as $\lim_{n\to\infty}\left(\dfrac{\ln(n+1)}{n+1}\times\dfrac{n}{\ln n}\right)=1.$

But by the integral test $\int_{1}^{\infty}\dfrac{\ln x}{x}\,dx=\left[\dfrac{(\ln x)^2}{2}\right]_{1}^{\infty}=\infty$ which is divergent meaning that

$\sum_{n=2}^{\infty}\dfrac{\ln n}{n}$ is also divergent.

EXAMPLE 4: Show that the series $\sum_{n=2}^{\infty}\left(\dfrac{1}{e}\right)^n$ converges and find its sum.

SOLUTION: The series $\sum_{n=2}^{\infty}\left(\dfrac{1}{e}\right)^n$ is geometric $\left(r=\dfrac{1}{e}<1\right)$ even though its first term is created with $n=2$ instead of $n=0$. To find its sum, we then calculate $\sum_{n=0}^{\infty}\left(\dfrac{1}{e}\right)^n=\dfrac{1}{1-\dfrac{1}{e}}=\dfrac{e}{e-1}$ and then subtract the first two terms, namely 1 and $\dfrac{1}{e}$. So $\sum_{n=2}^{\infty}\left(\dfrac{1}{e}\right)^n=\dfrac{e}{e-1}-1-\dfrac{1}{e}=\dfrac{1}{e^2-e}.$

EXAMPLE 5: Use partial fraction decomposition to show that $\displaystyle\sum_{n=4}^{\infty}\frac{2}{n^2-4n+3}$ converges and find its sum.

SOLUTION:

$$\sum_{n=4}^{\infty}\frac{2}{n^2-4n+3}=\sum_{n=4}^{\infty}\left(\frac{1}{n-3}-\frac{1}{n-1}\right)=\frac{1}{1}-\frac{1}{3}+\frac{1}{2}-\frac{1}{4}+\frac{1}{3}-\frac{1}{5}+\frac{1}{4}-\frac{1}{6}+...$$

All terms will cancel out except 1 and $\dfrac{1}{2}$ so the series converges to $\dfrac{3}{2}$.

EXAMPLE 6: Find the error in approximating $\displaystyle\sum_{n=1}^{\infty}\frac{\sin\left[\left(\dfrac{2n+1}{2}\right)\pi\right]}{\sqrt{n}}$ = using 8 terms.

SOLUTION: $\displaystyle\sum_{n=1}^{\infty}\frac{\sin\left[\left(\dfrac{2n+1}{2}\right)\pi\right]}{\sqrt{n}}=-\frac{1}{1}+\frac{1}{\sqrt{2}}-\frac{1}{\sqrt{3}}+...$ This is an alternating series with the absolute value of each term smaller than the one preceding it. Since it thus converges, the error in using 8 terms is the absolute value of the 9th term which is $\left|\dfrac{-1}{\sqrt{9}}\right|=\dfrac{1}{3}$.

EXAMPLE 7: For the following series, show work to determine whether they are convergent or divergent.

a) $\displaystyle\sum_{n=1}^{\infty}\frac{n+3}{2n-5}$

b) $\displaystyle 100\sum_{n=0}^{\infty}(0.99)^n$

c) $8-2+\dfrac{1}{2}-\dfrac{1}{8}+\dfrac{1}{32}-...$

d) $\displaystyle\sum_{n=1}^{\infty}\frac{5}{2n-1}$

e) $\displaystyle\sum_{n=1}^{\infty}\frac{1}{n^e}$

f) $\displaystyle\sum_{n=0}^{\infty}\frac{(-1)^n}{(2n)!}$

g) $\displaystyle\sum_{n=0}^{\infty}\frac{3}{2^n+1}$

h) $\displaystyle\sum_{n=2}^{\infty}\frac{1}{\sqrt{n^2-1}}$

i) $\displaystyle\sum_{n=0}^{\infty}\frac{9^{n+4}}{10^n}$

j) $\displaystyle\sum_{n=0}^{\infty}\left[(-1)^{n+1}\frac{n^5 4^{n+1}}{5^n}\right]$

k) $\displaystyle\sum_{n=1}^{\infty}\frac{n!}{n^n}$

l) $\displaystyle\sum_{n=1}^{\infty}ne^{-n}$

SOLUTIONS: a) Since $\displaystyle\lim_{n\to\infty}\frac{n+3}{2n-5}=\frac{1}{2}$, the series fails the nth-term test and is divergent.

b) This is a geometric series with $r = 0.99 < 1$ so it is convergent.

c) $8-2+\dfrac{1}{2}-\dfrac{1}{8}+\dfrac{1}{32}-...=8\left(1-\dfrac{1}{4}+\dfrac{1}{16}-\dfrac{1}{64}+...\right)=8\displaystyle\sum_{n=0}^{\infty}\left(-\dfrac{1}{4}\right)^n.$

This is geometric with $r=\dfrac{-1}{4}$ so it is convergent. The alternating test works as well.

d) This is a harmonic series and thus is divergent.

e) This is a p-series with $p =e > 1$ so it is convergent.

f) This is an alternating series that passes the nth-term test. Since $\dfrac{1}{(2n)!}>\dfrac{1}{(2n+2)!}$, the series converges.

g) The limit comparison test works well here. Comparing it to the convergent geometric series $\displaystyle\sum_{n=0}^{\infty}\frac{1}{2^n}$, we get $\displaystyle\lim_{n\to\infty}\left(\frac{3}{2^n+1}\times\frac{2^n}{1}\right)=3.$ Since 3 is finite and positive, the series converges.

h) Again, the limit comparison test works here.

Comparing it to divergent harmonic series $\displaystyle\int_{n=1}^{\infty} \frac{1}{n}$, we get

$$\int_{n=2}^{\infty} \left(\frac{1}{\sqrt{n^2-1}} \times \frac{n}{1} \right) = \int_{n=2}^{\infty} \left(\frac{n}{\sqrt{n^2-1}} \times \frac{1}{1} \right) = 1.$$ Since 1 is finite and positive, the series diverges.

i) By the ratio test, $\displaystyle\lim_{n\to\infty} \left| \frac{9^{n+5}}{10^{n+1}} \times \frac{10^n}{9^{n+4}} \right| = \frac{9}{10} < 1$ so the series is convergent.

j) It appears that the numerator is way more powerful than the denominator (and it is for 111 terms) but by the ratio test,

$$\lim_{n\to\infty} \left| (-1) \frac{(n+1)^5 \, 4^{n+2}}{5^{n+1}} \times \frac{5^n}{n^5 \, 4^{n+1}} \right| = \frac{4}{5} < 1,$$ the series is convergent.

k) Factorials usually do well with the ratio test.

$$\lim_{n\to\infty} \left| \frac{(n+1)!}{(n+1)^{n+1}} \times \frac{n^n}{n!} \right| = \lim_{n\to\infty} \left| (n+1) \times \frac{n^n}{(n+1)^n (n+1)} \right| = \lim_{n\to\infty} \left| \left(\frac{n}{n+1} \right)^n \right| = \frac{1}{e} < 1.$$

We leave it to an exercise for students to use L'Hospital's rule to show $\displaystyle\lim_{n\to\infty} \left| \left(\frac{n}{n+1} \right)^n \right| = \frac{1}{e}$.

So the series is convergent.

l) This problem can be done by the ratio test but the integral test also works: $\displaystyle\int_{1}^{\infty} xe^{-x} \, dx = \left[-xe^{-x} - e^{-x} \right]_{1}^{\infty} = 0$ by integration by parts and L'Hospital's rule. It is convergent.

TEST TIP

Infinite Series can be the most intimidating part of the BC exam. There is a lot of notation with the sigma notation itself, fractions, powers, and factorials. When you are given a series where the rule for the nth-term is given, a good piece of advice is to write out 3 or 4 terms of the series to get a feel for it. It makes the series less abstract and more understandable. Similarly, when you are given the first few terms of a series and need to generate the nth-term for a convergence test, look for patterns of coefficients, powers, and factorials. Remember the basic factorials for the integers one to six: 1, 2, 6, 24, 120, 720.

EXAMPLE 8: A solid known as Gabriel's Wedding Cake (shown partially below) is generated by rotating the function $f(x)$ about the x-axis where

$$f(x) = \begin{cases} 1 & \text{for } 1 \le x < 2 \\ \dfrac{1}{2} & \text{for } 2 \le x < 3 \\ \dfrac{1}{3} & \text{for } 3 \le x < 4 \\ \cdots \\ \dfrac{1}{n} & \text{for } n \le x < n+1 \\ \cdots \end{cases}$$

Show that the cake has finite volume.

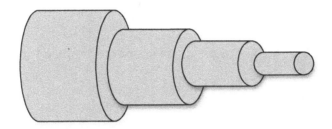

SOLUTION:
$$V = \sum_{n=1}^{\infty} \pi \left(\frac{1}{n} \right)^2 = \pi \sum_{n=1}^{\infty} \frac{1}{n^2}$$

This is a p-series with $p = 2 > 1$ that converges.

An interesting exercise is to show that cake has infinite surface area. This means that since the volume is convergent, we have a cake that we can eat but there isn't enough frosting in the world to cover it.

Power Series and Interval of Convergence

Overview: A **power series** is in the form of $\displaystyle\sum_{n=0}^{\infty} a_n x^n = a_0 + a_1 x + a_2 x^2 + a_3 x^3 + \dots$

Think of it as an infinite polynomial. This power series is said to be centered at $x = 0$. A power series in the form of

$$\sum_{n=0}^{\infty} a_n (x-c)^n = a_0 + a_1 (x-c) + a_2 (x-c)^2 + a_3 (x-c)^3 + \dots \text{ is centered at } x = c.$$

We are interested in values of x for which the power series converges. A power series will always converge at its center. The question is whether there are other values for which the series converges. Power series can converge at a) a single point, b) an interval (4 possible situations) or c) all real numbers.

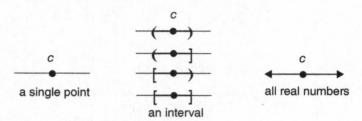

The **radius of convergence** represents the interval (the set of x-values) for which the power series converges and is denoted by R. If the series only converges at $x = c$, then $R = 0$. If the series converges for all real numbers, then $R = \infty$.

EXAMPLE 9: Find the interval of convergence for $\displaystyle\sum_{n=0}^{\infty} n! x^n$.

SOLUTION: $\displaystyle\sum_{n=0}^{\infty} n! x^n = 0! x^0 + 1! x^1 + 2! x^2 + 3! x^3 + \ldots = 1 + x + 2x^2 + 6x^3 + \ldots$
(Note: $0!$ is defined as 1.)

This is a power series centered at 0 and the series converges at 0 (if $x = 0$, the sum of the series is 1).

The question is: are there other values of x for which the series converges? To test this, we use a convergence test and the ratio test is best because of the factorials.

$\displaystyle\lim_{n\to\infty}\left|\frac{(n+1)! x^{n+1}}{n! x^n}\right| = \lim_{n\to\infty}\left|(n+1)x\right|$. The only value of x for which this expression is less than 1 is $x = 0$. So this series converges only at its center: $x = 0$. The radius of convergence $R = 0$.

EXAMPLE 10: Find the interval of convergence for $\sum_{n=0}^{\infty} (-1)^n \dfrac{x^n}{n!}$.

SOLUTION: $\sum_{n=0}^{\infty} (-1)^n \dfrac{x^n}{n!} = \dfrac{x^0}{0!} - \dfrac{x^1}{1!} + \dfrac{x^2}{2!} - \dfrac{x^3}{3!} + \ldots = 1 - x + \dfrac{x^2}{2} - \dfrac{x^3}{6} + \ldots$

This is a power series centered at 0 and the series converges at 0 (if $x = 0$, the sum of the series is 1).

The question is: are there other values of x for which the series converges? To test this, we use a convergence test and again the ratio test is best because of the factorials.

$\lim_{n \to \infty} \left| (-1) \dfrac{x^{n+1}}{(n+1)!} \times \dfrac{n!}{x^n} \right| = \lim_{n \to \infty} \left| \dfrac{-x}{n+1} \right|$. This converges for all values

of x as $\lim_{n \to \infty} \left| \dfrac{-x}{n+1} \right| = 0$ no matter the value of x. So this series converges for all real numbers and the interval of convergence is $(-\infty, \infty)$. This means that no matter the value of x, the series will always have a limit and never be infinite.

EXAMPLE 11: Find the interval of convergence for $\sum_{n=0}^{\infty} \dfrac{(x-3)^n}{n+1}$.

SOLUTION: $\sum_{n=0}^{\infty} \dfrac{(x-3)^n}{n+1} = 1 + \dfrac{(x-3)}{2} + \dfrac{(x-3)^2}{3} + \dfrac{(x-3)^3}{4} + \ldots$

This is a power series centered at 3 and the series converges at 3 (if $x = 3$, the sum of the series is 1).

The question is: are there other values of x for which the series converges? To test this, we use a convergence test and the ratio test is probably the best choice.

$\lim_{n \to \infty} \left| \dfrac{(x-3)^{n+1}}{n+2} \times \dfrac{n+1}{(x-3)^n} \right| = \lim_{n \to \infty} |x - 3| = |x - 3| < 1$

$x - 3 < 1$ so $x < 4$ $\quad\quad -(x - 3) < 1$ so $x > 2$

At this point, we can say that the interval of convergence is $(2, 4)$ and the radius of convergence R is 1. However, we have to check what happens at $x = 2$ and $x = 4$.

$x = 4$: the series is $1 + \dfrac{1}{2} + \dfrac{1}{3} + \dfrac{1}{4} + \ldots$ which is the divergent harmonic series.

$x = 2$: the series is $1 - \dfrac{1}{2} + \dfrac{1}{3} - \dfrac{1}{4} + \ldots$ which is the convergent alternating series.

So the interval of convergence is $[2, 4)$.

EXAMPLE 12: Find the interval of convergence for $\displaystyle\sum_{n=0}^{\infty} (4x)^n$.

SOLUTION: 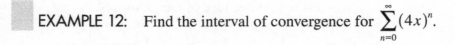 $\displaystyle\sum_{n=0}^{\infty} (4x)^n = 1 + 4x + 16x^2 + 64x^3 + \ldots$

This is a power series centered at 0 and the series converges at 0 (if $x = 0$, the sum of the series is 1).

The question is: are there other values of x for which the series converges? To test this, we use a convergence test and we will use a geometric test (although a ratio test works just as well).

This series is geometric and will converge when $|4x| < 1$ which gives $-\dfrac{1}{4} < x < \dfrac{1}{4}$.

At this point, we can say that the interval of convergence is $\left(-\dfrac{1}{4}, \dfrac{1}{4}\right)$ and the radius of convergence R is $\dfrac{1}{4}$. However, we have to check what happens at $x = -\dfrac{1}{4}$ and $x = \dfrac{1}{4}$.

$x = -\dfrac{1}{4}$: the series is $1 - 1 + 1 - 1\ldots$ which is divergent (*n*th-term test).

$x = \dfrac{1}{4}$: the series is $1 + 1 + 1 + 1\ldots$ which is divergent (*n*th-term test).

Thus the interval of convergence is $\left(-\dfrac{1}{4}, \dfrac{1}{4}\right)$.

EXAMPLE 13:   Find the interval of convergence for $\sum_{n=0}^{\infty}(-1)^n\dfrac{(3x+6)^n}{(2n)!}$.

SOLUTION: $\sum_{n=0}^{\infty}(-1)^n\dfrac{(3x+6)^n}{(2n)!}=\dfrac{1}{1}-\dfrac{(3x+6)}{2}+\dfrac{(3x+6)^2}{24}+...$

Don't be afraid by how complicated this looks. This is a power series centered at $x=-2$ and the series converges at -2 (if $x=-2$, the sum of the series is 1). The question is: are there other values of x for which the series converges? To test this, we use a convergence test and we will use the ratio test because of the factorials.

$$\lim_{n\to\infty}\left|(-1)\dfrac{(3x+6)^{n+1}}{(2n+2)!}\times\dfrac{(2n)!}{(3x+6)^n}\right|=\lim_{n\to\infty}\left|\dfrac{-(3x+6)}{(2n+2)(2n+1)}\right|=0$$

This is always convergent so the convergence doesn't depend on x. The interval of convergence is $(-\infty, \infty)$ meaning that no matter the value of x, the series will always have a limit and never be infinite. There is no need to test endpoints because there are no endpoints.

Taylor Polynomial Approximations

Overview: Taylor polynomial approximations allow students to write a polynomial that approximates a function. They are important because transcendental functions like trig functions or log functions can be expressed with nothing but addition and multiplication. For instance, when a student punches in $\sin 25°$, the calculator does nothing involving trig — it approximates the value by a series of additions and multiplications.

When we have a function $f(x)$ and some number c, we can approximate $f(x)$ using a polynomial. This is called the **nth degree Taylor polynomial approximation.**

The formula for the nth degree Taylor polynomial approximation for a function $f(x)$ at $x = c$ is given by:

$$P_n(x) = f(c) + f'(c)(x-c) + \frac{f''(c)(x-c)^2}{2!} + \frac{f'''(c)(x-c)^3}{3!} + \ldots + \frac{f^{(n)}(c)(x-c)^n}{n!}$$

If $c = 0$, then the Taylor polynomial is called a Maclaurin polynomial and is given by:

$$P_n(x) = f(0) + f'(0)x + \frac{f''(0)}{2!}x^2 + \frac{f'''(0)}{3!}x^3 + \ldots + \frac{f^{(n)}(0)}{n!}x^n$$

Typical AP-type problems have asked students to do one of the following:

- Given $f(x)$, find the nth-degree Taylor polynomial for $f(x)$ and use it to approximate $f(c)$.

- Given a table of values, find the nth-degree Taylor polynomial for $f(x)$ and use it to approximate $f(c)$.

- Given a formula for the nth derivative of a function, find the nth-degree Taylor polynomial for $f(x)$ and use it to approximate $f(c)$.

- Given the Taylor polynomial approximation of $f(x)$, find the values of specific derivatives at some value of c.

Taylor polynomial approximation problems go hand-in-hand with the second derivative test. If there is no first-degree x-term in the Taylor polynomial, then the value of c about which the function is centered is a critical value. Thus, the coefficient of x^2 is the second derivative divided by 2! Using the second derivative test, we can tell whether there is a relative maximum, minimum, or neither at $x = c$.

EXAMPLE 14: Find the 3ʳᵈ degree Maclaurin polynomial for $f(x) = e^{-2x}$ and use it to approximate $f(0.3)$.

SOLUTION:

$$f(x) = e^{-2x} \qquad f(0) = 1$$

$$f'(x) = -2e^{-2x} \qquad f'(0) = -2$$

$$f''(x) = 4e^{-2x} \qquad f''(0) = 4$$

$$f'''(x) = -8e^{-2x} \qquad f'''(0) = -8$$

$$f(x) \approx P_3(x) = 1 - 2x + \frac{4x^2}{2!} - \frac{8x^3}{3!} = 1 - 2x + 2x^2 - \frac{4x^3}{3}$$

$$f(0.3) \approx 1 - 2(0.3) + 2(0.3)^2 - \frac{4(0.3)^3}{3} = 1 - 0.6 + 0.18 - .036 = 0.544$$

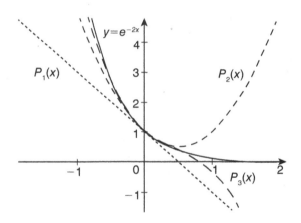

In the figure above, the function $y = e^{-2x}$ is shown in bold and $P_1(x)$, $P_2(x)$ and $P_3(x)$ are drawn. Note that the greater the number of terms in the Taylor polynomial, the better the approximation the polynomial is to the function.

EXAMPLE 15: Let f be a function having derivatives for all orders of real numbers. The first four derivatives of f at $x = 2$ are given in the table below. The function f and these four derivatives are increasing at $x = 2$.

x	$f(x)$	$f'(x)$	$f''(x)$	$f'''(x)$	$f^{(4)}(x)$
2	3	−2	4	6	−48

a) Write the first-degree Taylor polynomial for f about $x = 2$ and use it to approximate $f(1.9)$. Is this approximation greater than or less than $f(1.9)$? Explain your reasoning.

b) Write the fourth-degree Taylor polynomial for f about $x = 2$ and use it to approximate $f(1.9)$.

SOLUTIONS: a) $P_1(x) = 3 - 2(x-2) = -2x + 7$

$f(1.9) \approx P_1(1.9) = -2(1.9) + 7 = 3.2$

$P_1(1.9) < f(1.9)$ since f' is increasing (f concave up) at $x = 2$

b) $P_4(x) \approx f(x) = 3 - 2(x-2) + \dfrac{4(x-2)^2}{2} + \dfrac{6(x-2)^3}{3!} - \dfrac{48(x-2)^4}{4!}$

$P_4(1.9) \approx 3 - 2(-0.1) + 2(-0.1)^2 + (-0.1)^3 - 2(-0.1)^4$

$= 3 + 0.2 + 0.02 - 0.001 - 0.0002 = 3.2188$

EXAMPLE 16: The Maclaurin series for a certain function f converges to $f(x)$ for all x in the interval of convergence. The graph of f has a horizontal tangent line at the point $(0, 4)$. The nth derivative of f at $x = 0$ is given by $f^{(n)}(0) = \dfrac{(-1)^n n!}{2^n}$ for $n \geq 2$.

a) Write the 3rd-degree Maclaurin polynomial for $f(x)$.

b) Determine whether f has a relative maximum, relative minimum, or neither at $x = 0$. Justify your answer.

SOLUTIONS: $f(0) = 4$ and since there is a horizontal tangent at $(0,4)$, $f'(0) = 0$

a) $f''(0) = \dfrac{(-1)^2 \, 2!}{2^2} = \dfrac{1}{2}$, $f'''(0) = \dfrac{(-1)^3 \, 3!}{2^3} = \dfrac{-3}{4}$

$P_3(x) = 4 + 0x + \left(\dfrac{1}{2}\right)\left(\dfrac{x^2}{2!}\right) - \left(\dfrac{3}{4}\right)\left(\dfrac{x^3}{3!}\right) = 4 + \dfrac{x^2}{4} - \dfrac{x^3}{8}$

b) $f'(0) = 0$, $f''(0) = \dfrac{1}{2} > 0$, so by the

2nd derivative test, f has a relative minimum at $x = 0$

EXAMPLE 17: Consider the differential equation $\dfrac{dy}{dx} = 3x^2 - 2y - 5$. Let $f(x)$ be a particular solution to this differential equation with initial condition $f(-1) = -2$. Find the 2nd-degree Taylor polynomial for y about $x = -1$.

SOLUTION: $\dfrac{dy}{dx}\bigg|_{x=-1, y=-2} = 3(-1)^2 - 2(-2) - 5 = 2$

$\dfrac{d^2y}{dx^2} = 6x - 2\dfrac{dy}{dx} \Rightarrow \dfrac{d^2y}{dx^2}\bigg|_{x=-1, y=-2} = 6(-1) - 2(2) = -10$

$P_2(x) = -2 + 2(x+1) - \dfrac{10(x+1)^2}{2} = -2 + 2(x+1) - 5(x+1)^2$

EXAMPLE 18: Let $P(x) = (x - 5) - 9(x - 5)^3 + 15(x - 5)^4$ be the fourth-degree Taylor polynomial for the function f about $x = 5$. What is the value of $f'''(5)$?

SOLUTION: $\dfrac{f'''(5)}{3!} = -9 \Rightarrow f'''(5) = -54$. Don't fall into the trap here. It's not the coefficient of the 3rd term in the series that is important, it is the coefficient of the 3rd-degree term you need to work with.

Error Bounds

Overview: When the Taylor polynomial alternates in sign $(+, -, +, - \ldots)$ and converges to some limit, the error is easily found: it is simply the absolute value of the next term in the series as the terms get smaller and smaller. But when the Taylor polynomial doesn't alternate in sign, we need to find the **Lagrange error bound**.

For many students and teachers, the Lagrange error is confusing. Typically there is only one free-response question on the BC exam concerning Lagrange errors. When the nth degree Taylor polynomial is found for an approximation to a function $f(x)$ at $x = c$, there is an error – the difference between $f(c)$ and the result of the Taylor polynomial approximation. The Lagrange error bound gives the maximum error that using the Taylor polynomial approximation will yield.

An analogy goes like this: You are on the lot of a used-car dealer. You want to purchase a used car for exactly \$5,500. The salesman knows he has several cars that price, but is not sure of their exact location. So he directs you to the section of the lot where all cars cost between \$5,000 and \$6,000. You know that once in that section, any car you choose will be different from your target price of \$5,500 by at most \$500. This \$500 is the maximum error between what you want to pay and the true price.

The nth-degree Taylor polynomial approximation for a function $f(x)$ at $x = c$ is given by

$$P_n(x) = f(c) + f'(c)(x-c) + \frac{f''(c)(x-c)^2}{2!} + \frac{f'''(c)(x-c)^3}{3!} + \ldots + \frac{f^{(n)}(c)(x-c)^n}{n!}$$

Remember that this is an approximation for $f(x)$. Stated a different way:

$$f(x) = f(c) + f'(c)(x-c) + \frac{f''(c)(x-c)^2}{2!} + \frac{f'''(c)(x-c)^3}{3!} + \ldots + \frac{f^{(n)}(c)(x-c)^n}{n!} + R_n(x)$$

where $R_n(x)$ is the error that is created when using the Taylor polynomial.

This $R_n(x)$ is called the Lagrange form of the remainder and is found by the formula: $R_n(x) = \left| \frac{f^{(n+1)}(z)}{(n+1)!}(x-c)^{n+1} \right|$. The z is some value between x and c. You will not find z. For any nth-degree Taylor polynomial, you will need to find the maximum value of the $(n+1)^{st}$ derivative at any value between x and c to calculate the Lagrange error.

EXAM TIP

Lagrange errors are usually small, so more than the normal 3 decimal-place accuracy is required. It is recommended that you set your calculator to 5 decimal-place accuracy when you encounter calculator-active Lagrange error problems.

There are generally two types of problems in the AP exam that ask for the Lagrange error. They set up as follows:

- the maximum value of the $(n+1)^{st}$ derivative is given to you.

- the maximum value of the $(n+1)^{st}$ derivative is easily determined (usually trig functions).

EXAMPLE 19: The function f has derivatives of all orders for all real numbers and $f^{(4)}(x) = xe^x$. If the 3rd degree Taylor polynomial for f about $x = 1$ is used to approximate f on the interval $[1,2]$, what is the Lagrange error bound for the maximum error on the interval $[1,2]$?

SOLUTION: The maximum value of xe^x on $[1,2]$ occurs at $x = 2$ so the maximum value of the 4th derivative is $2e^2$.

So the Lagrange error for the 3rd-degree Taylor polynomial is
$$\frac{2e^2(2-1)^4}{4!} = \frac{e^2}{12}.$$

EXAMPLE 20: The second-degree Maclaurin polynomial for f is given by $P_2(x) = 10 - 5x + 9x^2$. The third derivative of f satisfies the inequality $\left| f^{(3)}(x) \right| \le 16$ for all x in the interval $[0,1]$. Find the Lagrange error bound for the approximation to $f\left(\frac{1}{2}\right)$.

SOLUTION: Lagrange error: $\dfrac{16}{3!}\left|\dfrac{1}{2}\right|^3 = \dfrac{16}{6}\left(\dfrac{1}{8}\right) = \dfrac{1}{3}.$

EXAMPLE 21: (Calculator Active) Let f be a function having derivatives for all orders of real numbers. The function and its first three derivatives of f at $x = 5$ are given in the table below. The fourth-derivative of f satisfies the inequality $\left| f^{(4)}(x) \right| \le 20$ for all x in the interval $[4,5]$ If $f(5.3)$ is approximated using the third-degree Taylor polynomial, find the upper bound of the approximation.

x	$f(x)$	$f'(x)$	$f''(x)$	$f'''(x)$
5	1	4	8	18

SOLUTION:

$$\text{Lagrange error} = \left| \frac{20(x-5)^4}{4!} \right| \qquad P_3(x) = 1 + 4(x-5) + \frac{8(x-5)^2}{2!} + \frac{18(x-5)^3}{3!}$$

So the maximum value of $f(5.3) = 1 + 4(0.3) + 4(0.3)^2 + 3(0.3)^3 + \left| \frac{20(0.3)^4}{24} \right| = 2.64775$

EXAMPLE 22: (Calculator Active) Find the Lagrange error in calculating $f\left(\dfrac{\pi}{4}\right)$ for the third degree Taylor polynomial for $f(x) = \sin x + \cos x$ about $x = 0$.

SOLUTION:

$f'(x) = \cos x - \sin x \qquad f''(x) = -\sin x - \cos x$

$f'''(x) = -\cos x + \sin x \qquad f^{(4)}(x) = \sin x + \cos x$

The maximum value $f^{(4)}(z) = \sin z + \cos z$ on $\left[0, \dfrac{\pi}{4} \right]$ occurs at $z = \dfrac{\pi}{4}$ and is $\dfrac{\sqrt{2}}{2} + \dfrac{\sqrt{2}}{2} = \sqrt{2}$

Lagrange error is $\dfrac{\sqrt{2}\left(\dfrac{\pi}{4}\right)^4}{4!} \approx 0.02242.$

Note: Since no justification was required, you can use a graph to ascertain that the maximum value of $\sin x + \cos x$ on $\left[0, \dfrac{\pi}{4}\right]$ occurs at $x = \dfrac{\pi}{4}$. Or function analysis techniques can be used.

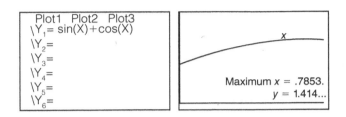

Taylor and Maclaurin Series

Overview: There is little distinction between Taylor series problems and Taylor polynomial problems other than the fact that the series is an infinite sum while the Taylor polynomial has a degree and stops at some value of n. Thus, the Taylor series is an exact sum, which converges to $f(x)$ while the Taylor polynomial is an approximation and will have an error associated with it.

Students should know the Taylor series for the following functions, the first three of which have an interval of convergence as $(-\infty, \infty)$. These crop up all the time.

$$\sin x = x - \frac{x^3}{3!} + \frac{x^5}{5!} - \frac{x^7}{7!} + \ldots + \frac{(-1)^n x^{2n+1}}{(2n+1)!} + \ldots$$

$$\cos x = 1 - \frac{x^2}{2!} + \frac{x^4}{4!} - \frac{x^6}{6!} + \ldots + \frac{(-1)^n x^{2n}}{(2n)!} + \ldots$$

$$e^x = 1 + x + \frac{x^2}{2!} + \frac{x^3}{3!} + \frac{x^4}{4!} + \ldots + \frac{x^n}{n!} + \ldots$$

$$\frac{1}{1-x} = 1 + x + x^2 + x^3 + \ldots + x^n + \ldots$$

Convergent on: $(-1, 1)$

Typical problems involve finding the radius or interval of convergence of Taylor series using the general term. Mostly, the ratio test is used.

EXAMPLE 23: What is the coefficient of x^4 in the Taylor series for $f(x) = \sqrt{4x+1}$ about $x = 0$?

SOLUTION:

$$f(x) = \sqrt{4x+1}$$

$$f'(x) = \frac{1}{2}(4x+1)^{-1/2}(4) = 2(4x+1)^{-1/2}$$

$$f''(x) = -(4x+1)^{-3/2}(4) = -4(4x+1)^{-3/2}$$

$$f'''(x) = 6(4x+1)^{-5/2}(4) = 24(4x+1)^{-5/2}$$

$$f^{(4)}(x) = -60(4x+1)^{-7/2}(4) = \frac{-240}{(4x+1)^{7/2}} \Rightarrow f^{(4)}(0) = -240$$

Coefficient of $x^4 = \dfrac{-240}{4!} = \dfrac{-240}{24} = -10$

EXAMPLE 24: The Maclaurin series for $f(x)$ is given by

$$\frac{1}{2!} - \frac{x}{3!} + \frac{x^2}{4!} - \frac{x^3}{5!} + \frac{x^4}{6!} \ldots + \frac{(-x)^n}{(n+2)!} + \ldots.$$

Find $f^{(18)}(0)$.

SOLUTION:

$$f'(x) = -\frac{1}{3!} + \frac{2x}{4!} - \frac{3x^2}{5!} + \frac{4x^3}{6!} + \ldots \Rightarrow f'(0) = -\frac{1!}{3!}$$

$$f''(x) = \frac{2}{4!} - \frac{6x}{5!} + \frac{12x^2}{6!} + \ldots \Rightarrow f''(0) = \frac{2!}{4!}$$

$$f'''(x) = -\frac{6}{5!} + \frac{24x}{6!} + \ldots \Rightarrow f'''(0) = -\frac{3!}{5!}$$

$$f^{(4)}(x) = \frac{24}{6!} + \ldots \Rightarrow f^{(4)}(0) = \frac{4!}{6!}$$

The pattern is established and $f^{(n)}(0) = (-1)^n \dfrac{n!}{(n+2)!} \Rightarrow f^{(18)}(0) = \dfrac{18!}{20!} = \dfrac{1}{380}$

EXAMPLE 25: The Maclaurin series for $f(x)$ is given by

$$\frac{x}{3!} - \frac{x^3}{5!} + \frac{x^5}{7!} - \frac{x^7}{9!} + \ldots + \frac{(-1)^n x^{2n+1}}{(2n+3)!} + \ldots$$

a) Let $g(x) = -x^2 f(x)$. Write the Maclaurin series for $g(x)$ showing the first three terms and the general term.

b) Write $g(x)$ in terms of a familiar function without using series. Using the same function, write $f(x)$.

SOLUTIONS: a) $g(x) = -x^2 f(x) = -\dfrac{x^3}{3!} + \dfrac{x^5}{5!} - \dfrac{x^7}{7!} + \ldots + \dfrac{(-1)^{n+1} x^{2n+3}}{(2n+3)!} + \ldots$

b) $g(x) = \sin x - x$

$$f(x) = \frac{g(x)}{-x^2} = \frac{\sin x - x}{-x^2} \quad \text{or} \quad f(x) = \frac{1}{x} - \frac{\sin x}{x^2}$$

EXAMPLE 26: If $f(x) = 2 - x + x^2 + \dfrac{x^3}{3!} + \dfrac{x^4}{4!} + \dfrac{x^5}{5!} + \ldots \dfrac{x^n}{n!} + \ldots$, find $\displaystyle\int_0^2 f(x)\, dx$.

SOLUTION: You have to recognize that $f(x)$ is similar to $e(x)$.

$$e^x = 1 + x + x^2 + \frac{x^3}{3!} + \frac{x^4}{4!} + \frac{x^5}{5!} + \ldots \frac{x^n}{n!} + \ldots$$

$$f(x) = e^x - 2x + 1 \Rightarrow \int_0^2 \left(e^x - 2x + 1\right) dx = \left[e^x - x^2 + x\right]_0^2$$

$$e^2 - 4 + 2 - (1 - 0 + 0) = e^2 - 3$$

Functions Defined as Power Series

Overview: Power series are similar to polynomials, but since they are series, they have an infinite number of terms.

A power series is in the form: $a_0 + a_1 x + a_2 x^2 + a_3 x^3 + \ldots + a_n x^n + \ldots$, centered at zero.

A power series in the form $a_0 + a_1(x-c) + a_2(x-c)^2 + a_3(x-c)^3 + \ldots + a_n(x-c)^n + \ldots$, is centered at c.

When you are given a formula for the nth-term of a series, write out the first 4 or 5 terms to see if it is in the form of a power series. Taylor series for a function $f(x)$ centered at c are special forms of power series where the coefficient of each term has the special relation: $a_n = \dfrac{f^{(n)}(c)}{n!}$. So all Taylor series are power series, but not all power series are Taylor series.

When you are given or asked to find a power series (or Taylor series) for some function $f(x)$, you are finding a polynomial with an infinite number of terms using the variable x. That allows you to find the value of the function at any value or variable by replacing that x with that number or variable. When you are asked to find the derivative or integral of $f(x)$, you can take the derivative or integral of the power series allowing you to possibly integrate an expression that might not otherwise be integrable.

For instance, if you were asked to write a Taylor series for $f(x) = e^{x^2}$ centered at 0, rather than go through the tedious process of taking derivatives, you use the fact that

$$f(x) = e^x = 1 + x + \frac{x^2}{2} + \frac{x^3}{3!} + \frac{x^4}{4!} + \frac{x^5}{5!} + \dots \text{ so}$$

$$f(x^2) = e^{x^2} = 1 + x^2 + \frac{(x^2)^2}{2} + \frac{(x^2)^3}{3!} + \frac{(x^2)^4}{4!} + \frac{(x^2)^5}{5!} + \dots.$$

You could find $\int e^{x^3}\, dx$ by integrating each term:

$$\int e^{x^2}\, dx = x + \frac{x^3}{3} + \frac{x^5}{5 \cdot 2!} + \frac{x^7}{7 \cdot 3!} + \frac{x^9}{9 \cdot 4!} + \dots + \frac{x^{2n+1}}{(2n+1) \cdot n!} + \dots + C.$$

EXAMPLE 27: Let $f(x)$ be the power series for cos x, centered at $x = 0$. Which of the following are power series?

a) $f(x^2)$

b) $f(\sqrt{x})$

c) $f\left(\dfrac{1}{x}\right)$

d) $f(e^x)$

SOLUTIONS:

$$f(x) = \cos x = 1 - \frac{x^2}{2!} + \frac{x^4}{4!} - \frac{x^6}{6!} + \ldots$$

a. $f(x^2) = \cos(x^2) = 1 - \dfrac{x^4}{2!} + \dfrac{x^8}{4!} - \dfrac{x^{12}}{6!} + \ldots$which is a power series.

b. $f(\sqrt{x}) = \cos(\sqrt{x}) = 1 - \dfrac{x}{2!} + \dfrac{x^2}{4!} - \dfrac{x^3}{6!} + \ldots$which is a power series.

c. $f\left(\dfrac{1}{x}\right) = \cos\left(\dfrac{1}{x}\right) = 1 - \dfrac{1}{x^2 2!} + \dfrac{1}{x^4 4!} - \dfrac{1}{x^6 6!} + \ldots$which is not a power series.

d. $f(e^x) = 1 - \dfrac{e^{2x}}{2!} + \dfrac{e^{4x}}{4!} - \dfrac{e^{6x}}{6!} + \ldots$which is not a power series.

EXAMPLE 28: The function f has derivatives of all orders for all real numbers x. Assume $f(-1) = -2, f'(-1) = 3, f''(1) = -2\ f'''(-1) = 6$.

a) Write the third-degree Taylor polynomial for f about $x = -1$ to approximate $f(-0.5)$.

b) Write the fourth-degree Taylor polynomial for $g(x) = f(x^2 - 1)$ about $x = -1$. Use your answer to explain why g must have a relative minimum at $x = 0$.

SOLUTIONS: a) $f(x) \approx P_3(x) = 2 + 3(x+1) - \dfrac{2(x+1)^2}{2} + \dfrac{6(x+1)^3}{3!}$

$f(-0.5) \approx 2 + 3(.5) - .5^2 + .5^3 = 2 + 1.5 - 0.25 + 0.125 = 3.375$

b) $g(x) \approx P_4 = f(x^2 - 1) = 2 + 3(x^2 - 1 + 1) - \dfrac{2(x^2 - 1 + 1)^2}{2}$

$= 2 + 3x^2 - x^4$

$g'(0) = 0$, $g''(0) = 6$. So, by the 2nd derivative test, g has a relative minimum at $x = 0$.

EXAMPLE 29: A power series is used to approximate $\displaystyle\int_0^1 e^{-\sqrt{x}}\, dx$ with a maximum error of 0.1. What is the minimum number of terms needed to obtain this approximation?

SOLUTION:

$e^x = 1 + x + \dfrac{x^2}{2!} + \dfrac{x^3}{3!} + \dfrac{x^4}{4!} + \ldots \Rightarrow e^{-\sqrt{x}} = 1 - x^{1/2} + \dfrac{x}{2!} - \dfrac{x^{3/2}}{3!} + \dfrac{x^2}{4!} + \ldots$

$\displaystyle\int_0^1 \left(1 - x^{1/2} + \dfrac{x}{2!} - \dfrac{x^{3/2}}{3!} + \dfrac{x^2}{4!} + \ldots \right) dx = \left[x - \dfrac{2x^{3/2}}{3} + \dfrac{x^2}{2 \cdot 2!} - \dfrac{2x^{5/2}}{5 \cdot 3!} + \dfrac{x^3}{3 \cdot 4!} + \ldots \right]_0^1$

$\displaystyle\int_0^1 e^{-\sqrt{x}}\, dx = 1 - \dfrac{2}{3} + \dfrac{1}{4} - \dfrac{1}{15} + \dfrac{1}{72} \ldots$

Since this is an alternating series and $\dfrac{1}{15} < 0.1$, 3 terms are needed.

DIDYOUKNOW?

π can be expressed as $4 - \dfrac{4}{3} + \dfrac{4}{5} - \dfrac{4}{7} + \ldots$ and we can show how. We know that

$\dfrac{1}{1+x} = 1 - x + x^2 - x^3 + \ldots$ so it follows that $\dfrac{1}{1+x^2} = 1 - x^2 + x^4 - x^6 + \ldots$ Integrating

both sides of that equation, we get $\tan^{-1} x = x - \dfrac{x^3}{3} + \dfrac{x^5}{5} - \dfrac{x^7}{7} + \ldots$ Plugging in 1, we

get $\tan^{-1} 1 = 1 - \dfrac{1}{3} + \dfrac{1}{5} - \dfrac{1}{7} + \ldots$ or $\dfrac{\pi}{4} = 1 - \dfrac{1}{3} + \dfrac{1}{5} - \dfrac{1}{7} +$ and $\pi = 4 - \dfrac{4}{3} + \dfrac{4}{5} - \dfrac{4}{7} + \ldots$

Unfortunately, this series converges slowly. It takes almost 300 terms to get even two-decimal-place accuracy.

EXAMPLE 30: Let f be the function given by $f(x) = \cos^{\sqrt{x}}$. Write the first 3 nonzero terms and the general term of the Taylor series for g about $x = 0$ where $g(x) = \dfrac{f(x) - 1}{x}$.

SOLUTION:

$$\cos x = 1 - \frac{x^2}{2!} + \frac{x^4}{4!} + \ldots + \frac{(-1)^n x^{2n}}{(2n)!} \qquad \cos^{\sqrt{x}} = 1 - \frac{x}{2!} + \frac{x^2}{4!} - \frac{x^3}{6!} \ldots + \frac{(-1)^n x^n}{(2n)!} + \ldots$$

$$g(x) = \frac{1 - \dfrac{x}{2!} + \dfrac{x^2}{4!} - \dfrac{x^3}{6!} \ldots + \dfrac{(-1)^n x^n}{(2n)!} + \ldots - 1}{x} = -\frac{1}{2!} + \frac{x}{4!} - \frac{x^2}{6!} + \ldots + \frac{(-1)^n x^{n-1}}{(2n)!} + \ldots, n \geq 1$$

EXAMPLE 31: (Calculator Active) The function shown below is $y = \dfrac{\sin 2x}{x^2}$. Find the difference in calculating the true area of the shaded region and by using the first three terms of a power series.

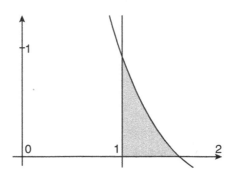

SOLUTION:

$$\sin x = x - \frac{x^3}{3!} + \frac{x^5}{5!} \ldots \Rightarrow \sin 2x = 2x - \frac{8x^3}{3!} + \frac{32x^5}{5!} - \ldots$$

$$\frac{\sin 2x}{x^2} = \frac{2}{x} - \frac{8x}{6} + \frac{32x^3}{120}$$

$$\sin 2x = 0 \Rightarrow x = \frac{\pi}{2}$$

$$\int_1^{\pi/2} \left(\frac{2}{x} - \frac{4x}{3} + \frac{4x^3}{15} \right) dx \approx 0.264 \qquad \int_1^{\pi/2} \left(\frac{\sin 2x}{x^2} \right) dx \approx 0.211$$

Difference $= 0.053$

EXAMPLE 32: The Maclaurin series for $\dfrac{1}{1-x}$ is $\displaystyle\sum_{n=0}^{\infty}(x)^n$. Find an infinite series

for $\dfrac{x}{1-\sqrt{x}}$.

SOLUTION:

$$f(x) = \frac{1}{1-x} = 1 + x + x^2 + x^3 + x^4 + \ldots$$

$$f(\sqrt{x}) = \frac{1}{1-\sqrt{x}} = 1 + x^{1/2} + x + x^{3/2} + x^3 + \ldots$$

$$\frac{x}{1-\sqrt{x}} = x\,f(\sqrt{x}) = x + x^{3/2} + x^2 + x^{5/2} + x^3 + \ldots$$

DIDYOUKNOW?

Did you know that the 5 most important numbers in all of mathematics are all linked by a single equation: $e^{i\pi} + 1 = 0$. In addition, we can prove it with your current knowledge: Using the fact that $e^x = 1 + x + \dfrac{x^2}{2!} + \dfrac{x^3}{3!} + \dfrac{x^4}{4!} + \dfrac{x^5}{5!} + \ldots$,

plug in $i\pi$ and we get $e^{i\pi} = 1 + i\pi + \dfrac{i^2\pi^2}{2!} + \dfrac{i^3\pi^3}{3!} + \dfrac{i^4\pi^4}{4!} + \dfrac{i^5\pi^5}{5!} + \ldots$. This

simplifies to $e^{i\pi} = 1 + i\pi - \dfrac{\pi^2}{2!} - \dfrac{i\pi^3}{3!} + \dfrac{\pi^4}{4!} + \dfrac{i\pi^5}{5!} + \ldots$. Rearranging, we get

$e^{i\pi} = 1 - \dfrac{\pi^2}{2!} + \dfrac{\pi^4}{4!} + \ldots + i\pi - \dfrac{i\pi^3}{3!} + \dfrac{i\pi^5}{5!} + \ldots$. Using Taylor expansions for sine

and cosine, we get $e^{i\pi} = \cos\pi + i\sin\pi = -1$. So, $e^{i\pi} = -1$ or $e^{i\pi} + 1 = 0$. This is known as Euler's Formula, one of the most remarkable formulas in all of math.

Time for a quiz
- Review strategies in Chapter 2
- Take Quiz 9 at the REA Study Center
  (www.rea.com/studycenter)

Take Mini-Test 3
on Chapters 9–11
Go to the REA Study Center
(www.rea.com/studycenter)

AP Calculus AB
Practice Exam

Also available at the REA Study Center *(www.rea.com/studycenter)*

This practice exam is available at the REA Study Center. Although AP exams are administered in paper-and-pencil format, we recommend that you take the online version of the practice exam for the benefits of:

- Instant scoring
- Enforced time conditions
- Detailed score report of your strengths and weaknesses

AP Calculus AB Practice Exam
Section I
Part A

(Answer sheets appear in the back of the book.)

TIME: 55 minutes

Number of Questions—28

Directions: Solve each of the following problems, select the best answer choice, and fill in the corresponding oval on the answer sheet.

Calculators may NOT be used for this section of the exam.

Notes:

(1) Unless otherwise specified, the domain of a function f is assumed to be the set of real numbers x for which $f(x)$ is a real number.

(2) The inverse of a trigonometric function may be indicated using the inverse notation f^{-1} or with the prefix "arc" (e.g., $\sin^{-1} x = \arcsin x$).

1. If $y = x(2 - 3x)^3$, then $\dfrac{dy}{dx} =$

   (A) $3x(2 - 3x)^2$

   (B) $-9(2 - 3x)^2$

   (C) $-9x(2 - 3x)^2$

   (D) $(2 - 3x)^2(1 - 9x)$

   (E) $-2(2 - 3x)^2(6x - 1)$

2. $\int (x^7 - 2x)^2 dx =$

(A) $\dfrac{x^3 (3x^{12} - 10x^6 + 60)}{45} + C$

(B) $\dfrac{x^3 (3x^{12} - 20x^6 + 60)}{45} + C$

(C) $\dfrac{(x^7 - 2x)^3}{3} + C$

(D) $\dfrac{x^{15}}{15} - \dfrac{4x^3}{3} + C$

(E) $\dfrac{x^{10}}{10} - \dfrac{4x^9}{9} + \dfrac{4x^3}{3} + C$

3. The graph of a differentiable function f is shown below. If $g(x) = \int_0^x f(t)\,dt$, which of the following is true?

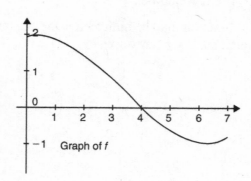

Graph of f

(A) $g''(4) < g(4) < g'(4)$

(B) $g(4) < g'(4) < g''(4)$

(C) $g(4) < g''(4) < g'(4)$

(D) $g'(4) < g''(4) < g(4)$

(E) $g''(4) < g'(4) < g(4)$

4. If $f'(x) = \cos x(x - 2)$, on what interval is $f(x)$ increasing if graphed on the interval $[0, 2\pi]$?

(A) $[0, 2\pi]$ only

(B) $\left[\dfrac{\pi}{2}, 2\right]$ and $\left[\dfrac{3\pi}{2}, 2\pi\right]$

(C) $\left[\dfrac{3\pi}{2}, 2\pi\right]$ only

(D) $[2, \pi]$ only

(E) $\left[\dfrac{\pi}{2}, 2\right]$ and $\left[2, \dfrac{3\pi}{2}\right]$

5. The table below gives selected values of $v(t)$, of a particle moving along the x-axis. At time $t = 0$, the particle is at the origin. Which of the following could be the graph of the position $x(t)$ of the particle for $0 \le t \le 4$?

t	0	1	2	3	4
$v(t)$	4	0	−0.5	−1	−1.5

(A)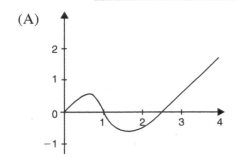

(D)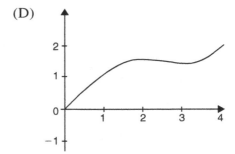

(B)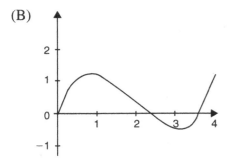

(E)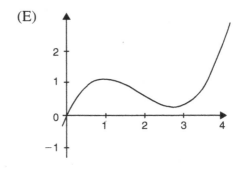

(C)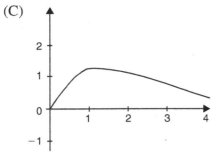

6. A swimming pool contains 1,000 gallons of water when it begins to drain at the end of the summer. Water is being drained from the pool at a rate of $R(t)$, where $R(t)$ is measured in gallons per hour. Selected values of $R(t)$ are given in the table below. Using a midpoint Riemann sum with two subintervals and data from the table, what is the approximate number of gallons of water in the tank at 5:00 PM?

t	9 AM	11 AM	1 PM	3 PM	5 PM
$R(t)$ gallons per hour	125	110	90	75	70

(A) 200

(B) 260

(C) 310

(D) 430

(E) 630

7. $\displaystyle\int_{1}^{e^2} \frac{(1+2\ln x)^2}{x}\, dx$ is equivalent to which of the following?

(A) $\displaystyle\frac{1}{2}\int_{1}^{5} u^2\, du$

(B) $\displaystyle 2\int_{1}^{5} u^2\, du$

(C) $\displaystyle\int_{1}^{3} u^2\, du$

(D) $\displaystyle 2\int_{1}^{3} u^2\, du$

(E) $\displaystyle\frac{1}{2}\int_{1}^{3} u^2\, du$

8. $\displaystyle\lim_{h \to 0} \frac{\tan^{-1}(1+h) - \tan^{-1}(1)}{h} =$

(A) 0

(B) $\dfrac{\pi}{4}$

(C) $\tan(1) \cdot \sec(1)$

(D) $\dfrac{1}{2}$

(E) nonexistent

9. If $f(x) = \begin{cases} \dfrac{a}{x} & \text{for } x \le 1 \\ 12 - bx^2 & \text{for } x > 1 \end{cases}$, find the values of a and b that make $f(x)$ differentiable.

   (A) $a = -24, b = 12$

   (B) $a = -8, b = 4$

   (C) $a = 8, b = -4$

   (D) $a = 4, b = -8$

   (E) no values make the function differentiable

10. A hot air balloon is at altitude 2000 feet. For the next 10 minutes it moves vertically and its velocity is given by the graph of the velocity $v(t)$ below where $v(t)$ is measured in ft/min. What is the altitude in feet of the balloon after 10 minutes?

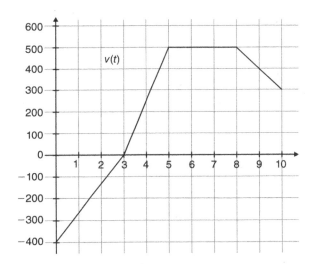

   (A) 1200

   (B) 1900

   (C) 3400

   (D) 4200

   (E) 5400

11. What are all the horizontal asymptotes of $f(x) = \dfrac{10 + 6(2^x)}{2 - 3(2^x)}$ in the xy-plane?

   (A) $y = -2$ only

   (B) $y = 2$ only

   (C) $y = 5$ only

   (D) $y = 0$ and $y = 2$ only

   (E) $y = -2$ and $y = 5$ only

12. If $x^2 + x - y^2 + y + 4 = 0$, find the behavior of the curve at $(-2, 3)$.

   (A) Increasing, concave up

   (B) Decreasing, concave up

   (C) Increasing, concave down

   (D) Decreasing, concave down

   (E) Decreasing, inflection point

13. If f is continuous for all real numbers x and $\displaystyle\int_0^3 f(x)\,dx = 8$, then $\displaystyle\int_2^5 [f(x-2) + 2x - 1]\,dx =$

   (A) 24

   (B) 26

   (C) 28

   (D) 42

   (E) not enough information

14. The graphs of $y = 4 - x^2$ and $y = 3x$ are shown in the figure below. Find the area of the shaded region.

   (A) $\dfrac{2}{3}$

   (B) $\dfrac{13}{6}$

   (C) $\dfrac{19}{6}$

   (D) $\dfrac{16}{3}$

   (E) $\dfrac{125}{6}$

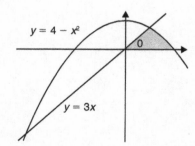

15. A tube of air is in the shape of a right circular cylinder. The height is increasing at the rate of 2.5 cm/sec while the radius is decreasing at the rate of 0.5 cm/sec. At the time that the radius is 2 cm and the height is 16 cm, how is the volume of the tube changing? (The area of a right circular cylinder with radius r and height h is $V = \pi r^2 h$.)

(A) Decreasing at $22\pi \ \dfrac{cm^3}{sec}$

(B) Increasing at $42\pi \ \dfrac{cm^3}{sec}$

(C) Decreasing at $80\pi \ \dfrac{cm^3}{sec}$

(D) Increasing at $39\pi \ \dfrac{cm^3}{sec}$

(E) Increasing at $64\pi \ \dfrac{cm^3}{sec}$

16. The function f is differentiable for all real numbers. The table below gives values of the function and its first derivatives at selected values of x. If f^{-1} is the inverse function of f, what is the equation for the line tangent to the graph of $y = f^{-1}(x)$ at $x = 5$?

x	$f(x)$	$f'(x)$
1	5	−2
5	−8	4

(A) $y+8 = \dfrac{1}{4}(x-5)$

(B) $y-1 = \dfrac{1}{4}(x-5)$

(C) $y-1 = -2(x-5)$

(D) $y+8 = \dfrac{-1}{2}(x-5)$

(E) $y-1 = -\dfrac{1}{2}(x-5)$

17. The graph of the piecewise linear function f (consisting of 3 lines) is shown in the figure below. If $g(x) = \int_{-1}^{x} f(t)\,dt$, arrange the values of these expressions from smallest to largest.

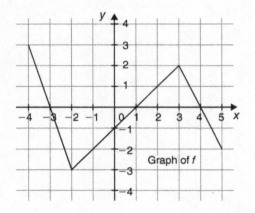

Graph of f

   I. $g(5)$
  II. $g(0)$
 III. $g(-4)$

(A) I, II, III

(B) I, III, II

(C) II, III, I

(D) II, I, III

(E) III, II, I

18. The velocity of a particle moving along the x-axis is given by the function $v(t) = \sin t \cdot e^{\cos t}$. What is the average velocity of the particle from time $t = \dfrac{\pi}{2}$ to $t = \pi$?

(A) $1 - \dfrac{1}{e}$

(B) $\dfrac{1}{e} - 1$

(C) $\dfrac{2(e-1)}{\pi e}$

(D) $\dfrac{2}{\pi e}$

(E) $\dfrac{2e}{\pi}$

19. An office copying machine is wireless and workers send jobs to print. Since pages are printed one at a time, the pages still to be printed are stacked up in what is called a queue. The number of pages in the queue at time t, measured in minutes, is modeled by a differentiable function Q where $0 \le t \le 8$. Values of $Q(t)$ are shown in the table below. For $0 \le t \le 8$, what is the fewest number of times in which $Q'(t) = 0$?

t (minutes)	0	1	2	3	4	5	6	7	8
$Q(t)$ pages	24	9	7	7	4	8	11	13	10

(A) 0

(B) 1

(C) 2

(D) 3

(E) 4

20. Let R be the region between the graphs of $y = \sqrt{x}$, $y = 3$, and the y-axis as shown in the figure below. The region R is the base of a solid with cross sections perpendicular to the x-axis as squares. Find the volume of the solid.

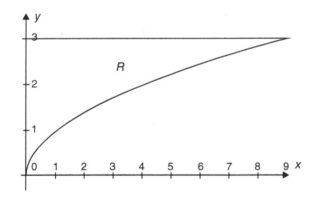

(A) 9

(B) $\dfrac{81}{2}$

(C) $\dfrac{27}{2}$

(D) 45

(E) $\dfrac{243}{2}$

21. The graph of f', the derivative of f, is shown below for $0 \le x \le 8$. The areas of the regions between the graph of f' and the x-axis are 8, 3 and 6, respectively. If $f(a) = 2$, what is the minimum value of f on the interval $0 \le x \le 8$?

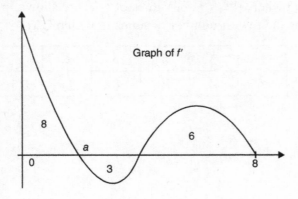

Graph of f'

(A) -8

(B) -6

(C) -3

(D) -1

(E) 1

22. A particle moving along a straight line that at any time $t > 0$ its velocity is given by $v(t) = \dfrac{\sin t - \cos t}{t}$. For which values of t is the particle slowing down?

I. $t = \dfrac{\pi}{6}$

II. $t = \dfrac{3\pi}{4}$

III. $t = \dfrac{3\pi}{2}$

(A) I only

(B) II only

(C) III only

(D) I and II only

(E) I and III only

23. The equation for the slope field below could be

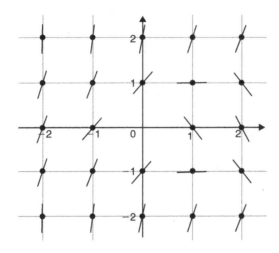

(A) $\dfrac{dy}{dx} = x + y^2$

(B) $\dfrac{dy}{dx} = x - y^2$

(C) $\dfrac{dy}{dx} = y^2 - x$

(D) $\dfrac{dy}{dx} = x^2 - y$

(E) $\dfrac{dy}{dx} = x + y$

24. The graph of the function f shown below has a horizontal tangent at $x = 1$. Let g be the continuous function defined by $g(x) = \int_{0}^{x} f(t)\ dt$. For what value(s) of x does the graph of g have a point of inflection?

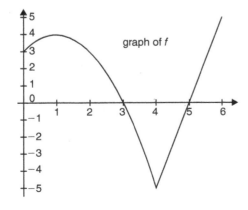

(A) 3 and 5

(B) 1 only

(C) 4 only

(D) 1 and 4

(E) no values

25. The difference in maximum acceleration and minimum acceleration attained on the interval $0 \leq t \leq 5$ by the particle whose velocity is given by $v(t) = 2t^3 - 6t^2 - 48t + 10$ is

(A) 104

(B) 96

(C) 54

(D) 48

(E) 42

26. Mike and Ike are laying bricks. Information about their work over the time interval $0 \leq t \leq 3$ is given in the table below. The number of bricks Mike lays is given by a differentiable function M of time t. The rate that Ike lays bricks is given by a differentiable function I of time t. Which of the following statements are necessarily true?

   I. At $t = 1.5$, Mike has laid 61 bricks.

   II. At $t = 1.5$, Ike's rate of brick laying is decreasing by approximately 16 bricks/hour.

   III. After 3 hours, Mike has laid more bricks than Ike.

t (hours)	0	1	2	3
Mike (bricks)	0	42	80	108
Ike (bricks per hour)	34	44	28	20

(A) I only

(B) II only

(C) III only

(D) I and II only

(E) II and III only

27. For the function f, $f'(x) = 3x - 5$ and $f(4) = 2$. What is the approximation for $f(3.9)$ found by using the tangent line to the graph of f at $x = 4$.

    (A) 1.3

    (B) 4.7

    (C) 6.7

    (D) 27.3

    (E) −2.7

28. Region R, enclosed by the graphs of $y = \pm 2\sqrt{x}$ and $y = 4 - 2x$ is shown in the figure below. Which of the following calculations would accurately compute the area of region R?

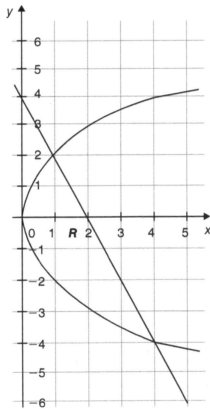

   I. $A = \int\limits_{0}^{1} 2\sqrt{x}\ dx + \int\limits_{1}^{4}(4 - 2x)\ dx$

   II. $A = \int\limits_{0}^{1} 4\sqrt{x}\ dx + \int\limits_{1}^{4}\left(4 - 2x + 2\sqrt{x}\right) dx$

   III. $A = \dfrac{1}{4}\int\limits_{-4}^{2}\left(8 - 2x - x^2\right) dx$

(A) I only

(B) II only

(C) III only

(D) II and III only

(E) I, II and III

STOP

This is the end of Section I, Part A.

If time still remains, you may check your work only in this section.

Do not begin Section I, Part B until instructed to do so.

Section I

Part B

TIME: 50 minutes

Number of Questions—17

Directions: Solve each of the following problems, select the best answer choice, and fill in the corresponding oval on the answer sheet.

A graphing calculator is required for some questions on this part of the exam.

Notes:

(1) The exact numerical value of the correct answer does not always appear among the choices given. When this happens, select from among the choices the number that best approximates the exact numerical value.

(2) Unless otherwise specified, the domain of a function f is assumed to be the set of real numbers x for which $f(x)$ is a real number.

(3) The inverse of a trigonometric function may be indicated using the inverse notation f^{-1} or with the prefix "arc" (e.g., $\sin^{-1} x = \arcsin x$).

76. A continuous function f is defined on the closed interval $-4 \le x \le 4$. The graph of the function, shown in the figure below consists of a line for $-4 \le x \le 0$ and then two curves, one on $0 \le x \le 2$ and the other on $2 \le x \le 4$. There is a value a, $-4 \le a < 4$, for which the Mean Value Theorem, applied to the interval $[a, 4]$ guarantees a value c, $a \le c < 4$ at which $f'(c) = -2$. What are possible values of a?

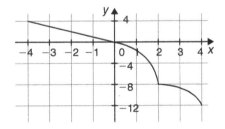

    I. -4

   II. 0

   III. 2

(A) I only

(B) II only

(C) III only

(D) I and III only

(E) none of these

77. The function f is defined for $x > 0$ with $f(\pi) = -1$, and $f'(x) = \cos\left(\dfrac{2}{x}\right)$. Use the line tangent to f at $x = \pi$ to approximate the value of $f(3)$ and whether it over or under-approximates $f(3)$.

(A) $-1.114 -$ under-estimates

(B) $-1.114 -$ over-estimates

(C) $0.886 -$ under-estimates

(D) $0.886 -$ over-estimates

(E) $0.804 -$ over-estimates

78. At time $t = 0$ minutes, ice cubes (32°F) are taken out of a freezer and placed in a bucket and taken outside where it is 85°F. The ice cube melts at the rate of $3e^{0.083t}$ degrees Fahrenheit per minute. To the nearest degree, what is the temperature of the melted water at time $t = 8$ minutes?

(A) 34°

(B) 38°

(C) 58°

(D) 66°

(E) 79°

79. If $y = \sec^{-1}(e^{2x})$, which of the following represents $\dfrac{dy}{dx}$?

(A) $2e^{2x} \sec(e^{2x}) \tan(e^{2x})$

(B) $\dfrac{-2e^{2x}}{\sin y}$

(C) $2e^{2x} \sin(e^{2x})$

(D) $2e^{2x} \cos^2(e^{2x})$

(E) $\dfrac{2e^{2x} \cos^2 y}{\sin y}$

80. Consider the differential equation $\dfrac{dy}{dx} = 4 - x^2 - 4y$. If $y = f(x)$ is the solution to the differential equation, at what point does f have a relative minimum?

(A) $(-1, -1)$

(B) $(-6, -8)$

(C) $(2, 0)$

(D) $(0, 1)$

(E) $(4, -3)$

81. The first-quadrant region bounded by $y = \cos x - x$, the x-axis, and the y-axis is rotated about the x-axis. Find the volume of the generated solid.

(A) 0.279

(B) 0.400

(C) 0.877

(D) 1.258

(E) 2.940

82. Find the average value of $f(x) = \dfrac{e^{\tan x}}{\cos^2 x}$ on the interval $\left[0, \dfrac{\pi}{4}\right]$.

(A) $\dfrac{4e}{\pi}$

(B) $\dfrac{4(e-1)}{\pi}$

(C) $4e$

(D) $\dfrac{e-1}{4\pi}$

(E) $\dfrac{e}{4\pi}$

83. Let f be a polynomial function with values of $f'(x)$ at selected values of x given in the table below. Which of the following must be true for $-3 < x < 6$?

x	-3	-1	1	3	4	6
$f'(x)$	-5	-8	7	2	1	3

(A) f is increasing

(B) f has at least two critical points

(C) f is concave down

(D) f has at least one inflection point

(E) the range of f is 8

84. A spherical balloon is being blown up. When its volume is 36π cm^3 and increasing at the rate of $20\pi \dfrac{\text{cm}^2}{\text{sec}}$, how fast is its diameter changing at that moment in $\dfrac{\text{cm}}{\text{sec}}$? (Volume of a sphere is $V = \dfrac{4}{3}\pi r^3$.)

(A) $\dfrac{5}{9}$

(B) $\dfrac{10}{9}$

(C) $\dfrac{5}{18}$

(D) 3

(E) $\dfrac{10}{3}$

85. The figure below shows the graph of $f(x)$. Which of the following statements are true?

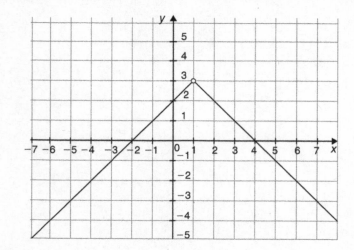

I. $\lim\limits_{x \to 1} f(x)$ exists

II. $\lim\limits_{x \to 1} f'(x)$ exists

III. $\lim\limits_{x \to 1} f''(x)$ exists

(A) I only

(B) II only

(C) III only

(D) I and II only

(E) I and III only

86. Jen tells her son Jeffrey to take a bath. Jeffrey fills the tub, but when the bathtub has 40 gallons of water in it, Jen decides to change the volume and temperature of the water for 6 minutes. The rate of change of the volume of the water is given by $v(t) = -0.8(t^3 - 6t^2 + t + 7)$, measured in gallons/min. Which of the following expressions gives the change in volume when the water is rising?

(A) $\displaystyle\int_{0}^{3.915} v(t)\ dt$

(B) $40 + \displaystyle\int_{0}^{3.915} v(t)\ dt$

(C) $\displaystyle\int_{1.337}^{5.598} v(t)\ dt$

(D) $\displaystyle\int_{1.337}^{5.598} v'(t)\ dt$

(E) $40 + \displaystyle\int_{1.337}^{5.598} v'(t)\ dt$

87. Find $\dfrac{d}{dx} \displaystyle\int_{0}^{\sin x} \dfrac{t}{1-t^2}\ dt.$

(A) $\tan x$

(B) $\dfrac{\sin x}{\cos^2 x}$

(C) $-\dfrac{1}{2}\ln(\cos^2 x)$

(D) $\sin x \cos x$

(E) $\dfrac{-1}{2\cos^2 x}$

88. The function f is continuous on the closed interval $[0, 7]$ and has the values given in the table below. The trapezoidal approximation for $\int_0^7 f(x)\ dx$ found with 3 subintervals is $15k$. What is the value of k?

x	0	2	5	7
f(x)	2	k²	5	8

(A) 3

(B) ±3

(C) 1

(D) 9

(E) no values of k

89. A particle moves along a line so that its acceleration for $t \geq 0$ is given by $a(t) = \dfrac{4t - 3}{e^{4t-3}}$. If the particle's velocity at $t = 1$ is 8.256, what was the velocity at $t = 0$?

(A) -10.227

(B) -2.227

(C) -1.971

(D) 18.483

(E) 21.044

90. The region R as shown in the figure below is bounded by the graphs of the equations $y = \cos x$, $y = 1$, and $x = \dfrac{\pi}{2}$. There is a value of k such that the area within R from $x = k$ to $x = \dfrac{\pi}{2}$ is $\dfrac{1}{2}$. To find k, which of the following equations needs to be solved?

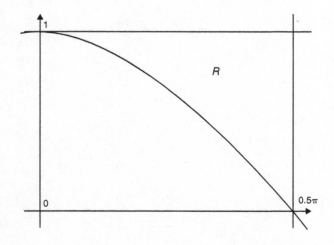

(A) $k - \sin k = \dfrac{\pi - 1}{2}$

(B) $k - \sin k = \dfrac{\pi - 3}{2}$

(C) $\sin k = \dfrac{\pi}{2} - 1$

(D) $\sin k = \dfrac{1}{2}$

(E) $k - \sin k = \dfrac{\pi + 1}{2}$

91. If $f'(x) = \sqrt{x^2 + 2x + 2} - x^2 + x - 2$, for what value of x does f have its absolute minimum on the interval [0, 2.5]?

(A) 0

(B) 0.440

(C) 1.168

(D) 1.444

(E) 2.5

92. A particle moves along the x-axis with position $x(t)$, velocity $v(t)$, and acceleration $a(t)$. Which of the following gives the average change of velocity on the time interval [0,10] ?

I. $\dfrac{v(10) - v(0)}{10}$

II. $\dfrac{x'(10) - x'(0)}{10}$

III. $\dfrac{1}{10} \displaystyle\int_0^{10} a(t)\, dt$

(A) I only

(B) II only

(C) III only

(D) I and II only

(E) I, II and III

STOP

This is the end of Section I, Part B.

If time still remains, you may check your work only in this section.

Do not begin Section II, Part A until instructed to do so.

Section II

Part A

Free-Response Questions

TIME: 30 minutes

    2 problems

Directions: Show all your work in you exam booklet. Grading is based on the methods used to solve the problems as well as the accuracy of your final answers.

A graphing calculator is required for this section.

Notes:

(1) Express your work in standard mathematical notation rather than calculator syntax.

(2) Unless otherwise specified, your final answers should be accurate to three decimal places.

(3) Unless otherwise specified, the domain of a function f is assumed to be the set of all real numbers x for which $f(x)$ is a real number.

Question 1

At closing time at a shopping mall, the mall does not magically empty. There are customers completing sales and employees who have to go through closing procedures. In the table below, $t = 0$ represents the announced closing time of the mall, negative numbers represent time in minutes before closing time and positive numbers represent time in minutes after closing time. The total number of people in the mall at time t is modeled by a strictly decreasing, twice-differentiable function P. Values of $P(t)$ at selected times t are given in the table below.

t (minutes)	-25	-15	0	10	30
$P(t)$ (people)	750	420	175	90	25

(a) Use the data in the table to estimate $P'(-5)$. Show the computation that leads to your answer. Using correct units, interpret the meaning of your answer in the context of this problem.

(b) Use the data in the table to evaluate $\int_{-25}^{30} P'(t)\, dt$. Using correct units, interpret the meaning of $\int_{-25}^{30} P'(t)\, dt$ in the context of this problem.

(c) For $-25 \le (t) \le 30$, approximate the average number of people in the mall by using a right Riemann sum with four subintervals. Does this approximation overestimate or underestimate the average number of people in the mall over this time period? Explain your reasoning.

(d) For $-25 \le (t) \le 30$, the function P that models the number of people in the mall has first derivative given by $P'(t) = -10.148(0.94)^t$. Based on the model, how many people were in the mall 5 minutes before closing time?

Question 2

Let R be the region in the first quadrant bounded by the x-axis and the graphs of $y = 2 \ln x$ and $y = 8 - \dfrac{x^2}{2}$ as shown in the figure below.

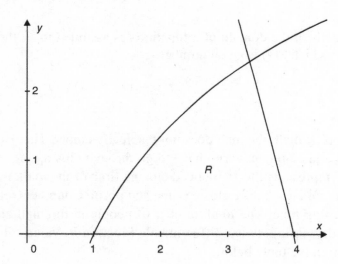

(a) Find the area of R.

(b) Region R is the base of a solid. For the solid, each cross section perpendicular to the x-axis is a square. Write, but do not evaluate an expression involving one or more integrals that gives the volume of this solid.

(c) The horizontal line $y = k$ divides R into two regions of equal area. Write but do not solve, and equation involving one or more integrals whose solution gives the value of k.

Section II

Part B

TIME: 60 minutes

    4 problems

Directions: Show all your work in your exam booklet. Grading is based on the methods used to solve the problems as well as the accuracy of your final answers.

No calculator is allowed for this section.

Question 3

Let f be the continuous function on $[-8, 6]$ whose graph consisting of three line segments, a semicircle of radius 1, and a semicircle of radius 2 is given at the figure below. Let g be the function given by $g(x) = \int_{-2}^{x} f(t)\, dt$.

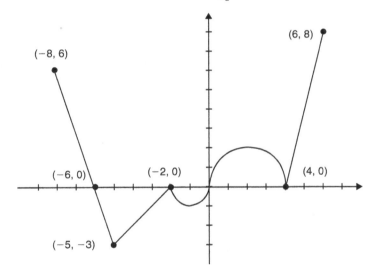

(a) Find $g(6)$ and $g(-6)$.

(b) For each of $g'(5)$ and $g''(5)$, find the value or state that it doesn't exist.

(c) Find the x-coordinate of each point at which the graph of g has a relative minimum or relative maximum. Justify your answers.

(d) What is the absolute maximum value of g?

(e) For $-8 < x < 6$, find all values of x for which the graph of g has an inflection point. Justify your answers.

Question 4

Newton the cat is mindlessly batting a piece of paper back and forth along the x-axis. Newton's velocity is modeled by a differentiable function v, where his position x is measured in meters and time t is measured in seconds. Selected values of $v(t)$ are given in the table below. Newton is at position $x = 6$ when $t = 0$ seconds.

t (seconds)	0	4	12	18	25	30
$v(t)$ (m/second)	5	8	-8	-4	-1	4

(a) Approximate Newton's acceleration at $t = 20$ seconds. Show your computation and indicate units of measure.

(b) Using a trapezoidal sum with five intervals, approximate Newton's position along the x-axis after 30 seconds.

(c) Using a trapezoidal sum with five intervals, approximate Newton's average speed over 30 seconds using correct units.

(d) Explain why Newton must have accelerated to the left at 2 m/sec at least once between $t = 4$ and $t = 12$ seconds.

(e) Suppose Newton's acceleration is positive for $0 \le t \le 4$ seconds. Find the minimum position of Newton along the x-axis at $t = 4$.

Question 5

Consider the closed curve in the xy-plane given by

$$y^3 - 9y + 4x^2 - 16x = 10$$

(a) Show that $\dfrac{dy}{dx} = \dfrac{16-8x}{3y^2-9}$.

(b) Find any x-coordinates where the curve has a horizontal tangent.

(c) Write an equation of the tangent line to the curve at the point $(4, -2)$ and use this line to approximate the value of y in the equation above when $x = 4.25$.

(d) Determine the concavity of the curve at the point $(4, -2)$.

(e) Is it possible for this curve to have a vertical tangent at points where the curve crosses the y-axis? Explain your reasoning.

Question 6

After a huge storm, a major city faced huge power failures and officials struggled to restore power. The rate at which power is restored is proportional to the difference between 100% and the percentage currently having power P. At time $t = 0$ immediately after the storm, only 25% of the city's population have power. If $P(t)$ is the percentage of people in the city with power t days after the storm ends, then

$$\frac{dP}{dt} = \frac{1}{3}(100 - P).$$

Let $y = P(t)$ be the solution to the differential equation above with initial condition $P(0) = 25$.

(a) Is power being restored faster when 40% of the city has power or when 60% of the city has power? Use appropriate units to explain your reasoning.

(b) Find $\dfrac{d^2P}{dt^2}$ in terms of P.

(c) Use $\dfrac{dP}{dt}$, $\dfrac{d^2P}{dt^2}$ and $\lim\limits_{p \to 100} \dfrac{dP}{dt}$ to draw a sketch of P on the graph below with $t \geq 0$.

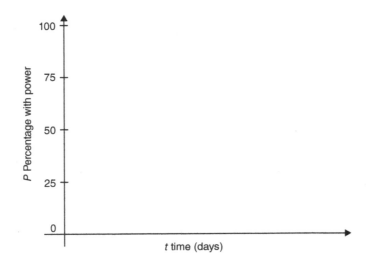

(d) Find the particular solution to find $P(t)$, the particular solution of the differential equation with initial condition $P(0) = 25$.

Answer Key

Section I

Part A

1. (E)	15. (A)		
2. (B)	16. (E)		
3. (E)	17. (D)		
4. (B)	18. (C)		
5. (C)	19. (D)		
6. (B)	20. (C)		
7. (A)	21. (B)		
8. (D)	22. (D)		
9. (B)	23. (C)		
10. (D)	24. (D)		
11. (E)	25. (B)		
12. (B)	26. (B)		
13. (B)	27. (A)		
14. (C)	28. (D)		

Part B

76. (C)	90. (B)
77. (A)	91. (E)
78. (D)	92. (E)
79. (E)	
80. (B)	
81. (C)	
82. (B)	
83. (D)	
84. (B)	
85. (E)	
86. (C)	
87. (A)	
88. (A)	
89. (D)	

Detailed Explanations of Answers

Section I, Part A

1. **(E)**

$$y' = x(3)(2 - 3x)^2 (-3) + (2 - 3x)^3$$
$$y' = -9x(2 - 3x)^2 + (2 - 3x)^3$$
$$y' = (2 - 3x)^2(-9x + 2 - 3x)$$
$$y' = (2 - 3x)^2(-12x + 2)$$
$$y' = -2(2 - 3x)^2 (6x - 1)$$

2. **(B)**

$$\int\left(x^{14} - 4x^8 + 4x^2\right) dx = \frac{x^{15}}{15} - \frac{4x^9}{9} + \frac{4x^3}{3}$$
$$= \frac{3x^{15} - 20x^9 + 60x^3}{45}$$
$$= \frac{x^3\left(3x^{12} - 20x^6 + 60\right)}{45} + C$$

3. **(E)**

$$g(4) = \int_0^4 f(t)\, dt \text{ which is the area under } f \text{ on } [\,0, 4] \text{ which is positive.}$$

$$g'(x) = \frac{d}{dx}\left[\int_0^x f(t)\, dt\right] = f(x) \text{ so } g'(4) = f(4) = 0$$

$g''(x) = f'(x)$ so $g''(4) = f'(4)$ which is the slope of f at $x = 4$ which is negative.

So $g''(4) < g'(4) < g(4)$.

4. **(B)**

$$\cos x(x-2) = 0 \Rightarrow x = \frac{\pi}{2}, \frac{3\pi}{2}, 2$$

$f'(x):$ $------0++++0--------0+++++++$

$0 \qquad \frac{\pi}{2} \qquad 2 \qquad\qquad \frac{3\pi}{2} \qquad\qquad 2\pi$

f increasing on $\left[\dfrac{\pi}{2}, 2\right]$ and $\left[\dfrac{3\pi}{2}, 2\pi\right]$

5. **(C)**

Velocity is zero at $t = 1$ so the graph of $x\,(t)$ must have horizontal tangents at $t = 1$ which are B, C, and E. The slope of the graph will be negative at $t = 2,3,4$ which is only C.

6. **(B)**

Gallons drained: $4(110) + 4(75) = 740$
Gallons remaining: $1000 - 740 = 260$

7. **(A)**

$$\int_1^{e^2} \frac{(1+2\ln x)^2}{x}\, dx \qquad u = 1+2\ln x, du = \frac{2}{x}dx$$

$$x = 1, u = 1, x = e^2, u = 1+2\ln e^2 = 1+4 = 5$$

$$\frac{1}{2}\int_1^{e^2} \frac{2(1+2\ln x)^2}{x}\, dx$$

$$\frac{1}{2}\int_1^5 u^2\, du$$

8. **(D)**

This is asking for the derivative of $y = \tan^{-1}(x)$ at $x = 1$.

$$y' = \frac{1}{1+x^2} \Rightarrow y'(1) = \frac{1}{1+x^2} = \frac{1}{2}.$$

9. **(B)**

continuity: $\lim\limits_{x \to -1^-} f(x) = -a \quad \lim\limits_{x \to -1^+} f(x) = 12 - b$ so $-a = 12 - b$

$$f'(x) = \begin{cases} \dfrac{-a}{x^2} \text{ for } x \le -1 \\ -2bx \text{ for } x > -1 \end{cases}$$

differentiability: $\lim\limits_{x \to -1^-} f'(x) = -a \quad \lim\limits_{x \to -1^+} f'(x) = 2b$ so $-a = 2b$

$2b = 12 - b \Rightarrow b = 4$ and $a = -8$.

10. **(D)**

$$\text{Position} = 2000 + \int_0^{10} v(t)\, dt$$

$$= 2000 - \frac{1}{2}(3)(400) + \frac{1}{2}(2)(500) + 3(500) + \frac{1}{2}(2)(500 + 300)$$

$$= 2000 - 600 + 500 + 1500 + 800 = 4200 \text{ ft.}$$

11. **(E)**

$$\lim_{x \to -\infty} \frac{10 + 6(2^x)}{2 - 3(2^x)} = \lim_{x \to \infty} \frac{10 + \dfrac{6}{2^x}}{2 - \dfrac{3}{2^x}} = 5 \qquad\qquad \lim_{x \to \infty} \frac{10 + 6(2^x)}{2 - 3(2^x)} = -2$$

12. **(B)**

$$2x + 1 - 2y\frac{dy}{dx} + \frac{dy}{dx} = 0$$

$$\frac{dy}{dx} = \frac{2x+1}{2y-1} \Rightarrow \frac{dy}{dx}\bigg|_{(-2,3)} = \frac{-4+1}{6-1} = \frac{-3}{5} < 0, \text{ so the curve is decreasing.}$$

$$\frac{d^2y}{dx^2} = \frac{2(2y-1) - 2(2x+1)\dfrac{dy}{dx}}{(2y-1)^2}$$

$$\frac{d^2y}{dx^2}\bigg|_{(-2,3)} = \frac{2(5) - 2(-3)\left(\dfrac{-3}{5}\right)}{5^2} = \frac{50 - 18}{125} = \frac{32}{25} > 0, \text{ so the curve is concave up.}$$

13. **(B)**

$$\int_2^5 [f(x-2)+2x-1]\,dx = \int_2^5 f(x-2)\,dx + \int_2^5 (2x-1)\,dx$$

$u = x - 2,\ du = dx\quad x = 2,\ u = 0,\ x = 5,\ u = 3$

$$\int_2^5 f(x-2)\,dx = \int_0^3 f(x)\,dx = 8 \qquad \int_2^5 (2x-1)\,dx = \left[x^2 - x\right]_2^5 = 25 - 5 - (4-2) = 18$$

$8 + 18 = 26$

14. **(C)**

$4 - x^2 = 0 \Rightarrow x = \pm 2$

$4 - x^2 = 3x \Rightarrow x^2 + 3x - 4 = 0 \Rightarrow (x-1)(x+4) = 0 \Rightarrow x = 1, -4$

$$A = \int_0^1 3x\,dx + \int_1^2 (4-x^2)\,dx$$

$$A = \left[\frac{3x^2}{2}\right]_0^1 + \left[4x - \frac{x^3}{3}\right]_1^2 = \frac{3}{2} + 8 - \frac{8}{3} - \left(4 - \frac{1}{3}\right) = \frac{3}{2} + 4 - \frac{7}{3} = \frac{9 + 24 - 14}{6} = \frac{19}{6}$$

Alternatively: $A = \int_0^3 \left(\sqrt{4-y} - \frac{y}{3}\right)dy = \frac{19}{6}$

15. **(A)**

$$\frac{dV}{dt} = \pi\left(r^2 \frac{dh}{dt} + 2rh \frac{dr}{dt}\right)$$

$$\frac{dV}{dt} = \pi\left[4(2.5) + 2(2)(16)\left(\frac{-1}{2}\right)\right]$$

$$\frac{dV}{dt} = \pi(10 - 32) = -22\pi$$

16. **(E)**

$f(1) = 5 \Rightarrow f^{-1}(5) = 1$

$$\left(f^{-1}\right)'(5) = \frac{1}{f'\left(f^{-1}(5)\right)} = \frac{1}{f'(1)} = \frac{1}{-2}$$

Tangent line : $y - 1 = -\frac{1}{2}(x - 5)$

17. **(D)**

$$g(5) = \int_{-1}^{5} f(t)\,dt = \int_{-1}^{1} f(t)\,dt + \int_{1}^{4} f(t)\,dt + \int_{4}^{5} f(t)\,dt = \frac{1}{2}(2)(-2) + \frac{1}{2}(3)(2) + \frac{1}{2}(1)(-2) = -2+3-1 = 0$$

$$g(0) = \int_{-1}^{0} f(t)\,dt = -1.5$$

$$g(-4) = \int_{-1}^{-4} f(t)\,dt = \int_{-1}^{-2} f(t)\,dt + \int_{-2}^{-3} f(t)\,dt + \int_{-3}^{-4} f(t)\,dt = 2.5 + \frac{1}{2}(-1)(-3) + \frac{1}{2}(-1)(3) = 2.5$$

18. **(C)**

$$v_{avg} = \frac{\displaystyle\int_{\pi/2}^{\pi} \sin t \cdot e^{\cos t}\,dt}{\pi - \dfrac{\pi}{2}} \qquad u = \cos t,\, du = -\sin t\,dt \qquad t = \frac{\pi}{2},\, u = 0 \quad t = \pi,\, u = -1$$

$$v_{avg} = \frac{-\displaystyle\int_{\pi/2}^{\pi} -\sin t \cdot e^{\cos t}\,dt}{\dfrac{\pi}{2}} = \frac{-\displaystyle\int_{0}^{-1} e^{u}\,du}{\dfrac{\pi}{2}} = \frac{-e^{u}\big|_{0}^{-1}}{\dfrac{\pi}{2}} = \frac{1 - \dfrac{1}{e}}{\dfrac{\pi}{2}} = \frac{2(e-1)}{\pi e}$$

19. **(D)**

Q is differentiable on [0,8] so Q is also continuous. Using the IVT and MVT, Q' must equal 0 at least once on (2,3),(3,5) and (6,8).

20. **(C)**

$$s = 3 - \sqrt{x}$$

$$A = s^2 = \left(3 - \sqrt{x}\right)^2 = 9 - 6x^{1/2} + x$$

$$V = \int_{0}^{9}\left(9 - 6x^{1/2} + x\right)\,dx = \left[9x - 6\left(\frac{2}{3}\right)x^{3/2} + \frac{x^2}{2}\right]_{0}^{9}$$

$$V = 81 - 4(27) + \frac{81}{2} = \frac{162 - 216 + 81}{2} = \frac{27}{2}$$

21. **(B)**

$$f(x) = f(a) + \int_0^x f'(t)\, dt.$$

$$f(0) = 2 + \int_a^0 f'(t)\, dt = 2 - 8 = -6$$

Call the next zero of $f' = b$: $f(b) = 2 + \int_a^b f'(t)\, dt = 2 - 3 = -1$

$$f(8) = 2 + \int_a^8 f'(t)\, dt = 2 - 3 + 6 = 5.$$

The minimum value of f on $[0, 8]$ is -6.

22. **(D)**

$$a(t) = \frac{t(\cos t + \sin t) - (\sin t - \cos t)}{t^2}$$

t	$v(t)$	$a(t)$
$\dfrac{\pi}{6}$	$\dfrac{\dfrac{1}{2} - \dfrac{\sqrt{3}}{2}}{\dfrac{\pi}{6}} < 0$	$\dfrac{\dfrac{\pi}{6}\left(\dfrac{\sqrt{3}}{2} + \dfrac{1}{2}\right) - \left(\dfrac{1}{2} - \dfrac{\sqrt{3}}{2}\right)}{\dfrac{\pi^2}{36}} > 0$
$\dfrac{3\pi}{4}$	$\dfrac{\dfrac{\sqrt{2}}{2} - \left(\dfrac{-\sqrt{2}}{2}\right)}{\dfrac{3\pi}{4}} > 0$	$\dfrac{\dfrac{3\pi}{4}\left(\dfrac{-\sqrt{2}}{2} + \dfrac{\sqrt{2}}{2}\right) - \left(\dfrac{\sqrt{2}}{2} + \dfrac{\sqrt{2}}{2}\right)}{\dfrac{9\pi^2}{16}} < 0$
$\dfrac{3\pi}{2}$	$\dfrac{-1 - 0}{\dfrac{3\pi}{2}} < 0$	$\dfrac{\dfrac{3\pi}{2}(0-1) - (-1-0)}{\dfrac{9\pi^2}{4}} < 0$

Particle slows down when v and a have opposite signs: $t = \dfrac{\pi}{6}$ and $t = \dfrac{3\pi}{4}$.

Note that it is not necessary to compute the values, just the signs.

23. **(C)**

Since the slope is 0 at $(1, 1)$, choices A and E are eliminated.
Since the slope is 0 at $(1, -1)$, choice D is eliminated
At $(0, 1)$ the slope field appears to be 1. That eliminates choice B.
Note that the slope field is symmetric with the x-axis. That is an extra confirmation of choice C.

24. **(D)**

$$g(x) = \int_0^x f(t)\, dt \Rightarrow g'(x) = f(x) \Rightarrow g''(x) = f'(x)$$
$$\qquad\qquad\quad \cup \qquad\qquad \cap \qquad\qquad \cup$$

$g''(x): \underline{+++++0--------\infty+++++}$ g changes concavity at $x = 1$, $x = 4$.
$\qquad\qquad 0 \qquad 1 \qquad\qquad 4 \qquad 6$

25. **(B)**

$a(t) = 6t^2 - 12t - 48 \Rightarrow a'(t) = 12t - 12 = 0 \Rightarrow t = 1$
$a(0) = -48, \quad a(1) = -54, \quad a(5) = 42$
Maximum acceleration = 42. Minimum acceleration = -54. Range = 96.

26. **(B)**

I is not necessarily true. We do not know that Mike is working at a steady rate.

II is true. $I'(t) \approx \dfrac{28 - 44}{2 - 1} = -16$.

III is not necessarily true. Using a left Riemann sum, Ike lays 106 bricks and using a right Riemann sum, Ike lays 92 bricks—which are both less than the 108 bricks that Mike lays. But since we do not know what is occurring in intermediate values, we can only approximate the number of bricks that Ike lays.

27. **(A)**

$f'(4) = 3(4) - 5 = 7$
Tangent line: $y - 2 = 7(x - 4) \Rightarrow y = 7x - 26$
$y(3.9) \approx 7(3.9) - 26 = 1.3$.
Be careful of the trap answer (E). You aren't given the tangent line equation, just the formula for the slope of the tangent line.

28. **(D)**

I does not take into consideration $y = -2\sqrt{x}$

II is accurate. $y = 2\sqrt{x}$ is the top curve and $y = -2\sqrt{x}$ is the bottom curve on $[0, 1]$ and $y = 4 - 2x$ is the bottom curve on $[1, 4]$

III is accurate. If we integrate with respect to y, we would use right minus left.

Right: If $y = 4 - 2x \Rightarrow x = \dfrac{4 - y}{2}$ Left: If $y = \pm 2\sqrt{x}$, then $x = \dfrac{y^2}{4}$.

That gives $\displaystyle\int_{-4}^{2} \left(\dfrac{4 - y}{2} - \dfrac{y^2}{4} \right) dy = \dfrac{1}{4} \int_{-4}^{2} \left(8 - 2y - y^2 \right) dy.$

y is a dummy variable, you can use any variable. (When you calculate it with your calculator, you use x.)

Detailed Explanations of Answers

Section I, Part B

76. **(C)**

The issue here is differentiability. The function is not differentiable at $x = 0$ and $x = 2$ so a cannot equal -4 or 0. The function is differentiable on $(2, 4)$ and $\dfrac{f(4) - f(2)}{4 - 2} = \dfrac{-12 + 8}{2} = -2$. So the MVT holds on $(2, 4)$ and $a = 2$.

77. **(A)**

Tangent line $f'(\pi) = \cos\left(\dfrac{2}{\pi}\right) = 0.804$

$y + 1 = 0.804(x - \pi) \Rightarrow y = 0.804(x - \pi) - 1 \Rightarrow y(3) \approx 0.804(3 - \pi) - 1 = -1.114$

$f''(x) = \left(\dfrac{2}{x^2}\right)\sin\left(\dfrac{2}{x}\right)$ $f''(x) > 0$ on $(3, \infty)$ so f is concave up and -1.114 is an underestimation.

78. **(D)**

$T(8) - T(0) = \displaystyle\int_0^8 3e^{0.083t}\, dt \Rightarrow T(8) = T(0) + \displaystyle\int_0^8 3e^{0.083t}\, dt \approx 66°$

```
32+fnInt (3e^(.083X),
X, 0, 8)
      66.06796395
```

79. **(E)**

$$\sec y = e^{2x}$$

$$\frac{1}{\cos y} = e^{2x}$$

$$\frac{\sin y}{\cos^2 y}\frac{dy}{dx} = 2e^{2x}$$

$$\frac{dy}{dx} = \frac{2e^{2x}\cos^2 y}{\sin y}$$

80. **(B)**

$$\frac{d^2y}{dx^2} = -2x - 4\frac{dy}{dx}$$

To have a relative minimum, $\dfrac{dy}{dx} = 0$ and $\dfrac{d^2y}{dx^2} > 0$.

A. $\dfrac{dy}{dx} = 7$

B. $\dfrac{dy}{dx} = 0$, $\dfrac{d^2y}{dx^2} = 12$

C. $\dfrac{dy}{dx} = 0$, $\dfrac{d^2y}{dx^2} = -4$

D. $\dfrac{dy}{dx} = 0$, $\dfrac{d^2y}{dx^2} = 0$

E. $\dfrac{dy}{dx} = 0$, $\dfrac{d^2y}{dx^2} = -8$

81. **(C)**

$$V = \pi \int_0^{0.739} (\cos x - x)^2 \, dx$$

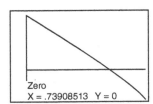

Zero
X = .73908513   Y = 0

πfnInt (Y1², X, 0, .739)
                .8770121901

82. **(B)**

$$f_{avg} = \frac{\int_0^{\pi/4} e^{\tan x} \sec^2 x \, dx}{\frac{\pi}{4} - 0} \qquad u = \tan x, du = \sec^2 x \, dx$$

$$x = 0, u = 0 \quad x = \frac{\pi}{4}, u = 1$$

$$f_{avg} = \frac{\int_0^1 e^u \, du}{\frac{\pi}{4}} = \frac{[e^u]_0^1}{\frac{\pi}{4}}$$

$$f_{avg} = \frac{4(e-1)}{\pi} \qquad \text{OR}$$

| fnInt (e^(tan (X))/(cos (X))², |
| X, 0, π/4)/(π/4) |
| 2.187784373 |
| 4(e−1)/π |
| 2.187784373 |

83. **(D)**

A. Since there are values for which $f'(x) < 0$, f is not always increasing.

B. $f'(x)$ switches from negative to positive between -1 and 1. Since f is a polynomial, it is continuous and differentiable and its derivatives are continuous. Thus, by the intermediate value theorem, there must be a value on $(-1, 1)$ where $f'(x) = 0$. But that is the only interval where $f'(x)$ must equal zero.

C. Graphs are concave down where $f''(x) < 0$ or when $f(x)$ is decreasing. By the argument in B, there must be a value on $(-1, 1)$ where $f(x) = 0$ and by the MVT, there must be a value c for which $f''(x) > 0$ on $(-11, c)$.

D. Since there must be a value a on $(3, 4)$ where $f''(a) = -1$ and there must be a value b on $(4, 6)$ where $f''(b) = 1$, f must change concavity at some value on $(3, 6)$.

E. Since we do not know intermediate values, we have no idea what the absolute maximum and absolute minimum values of f are.

84. **(B)**

$$36\pi = \frac{4}{3}\pi r^3 \Rightarrow r^3 = 27 \Rightarrow r = 3.$$

$$\frac{dV}{dt} = 4\pi r^2 \frac{dr}{dt}$$

$$20\pi = 4\pi(9)\frac{dr}{dt}$$

$$\frac{dr}{dt} = \frac{20\pi}{36\pi} = \frac{5}{9} \text{ so } \frac{d(\text{diam})}{dt} = 2\left(\frac{5}{9}\right) = \frac{10}{9}$$

85. **(E)**

$$f(x)=\begin{cases} x+2, x<1 \\ -x+4, x>1 \end{cases} \quad f'(x)=\begin{cases} 1, x<1 \\ -1, x>1 \end{cases} \quad f''(x)=\begin{cases} 0, x<1 \\ 0, x>1 \end{cases}$$

$$\lim_{x\to1^-} f(x)=3 \qquad \lim_{x\to1^-} f'(x)=1 \qquad \lim_{x\to1^-} f''(x)=0$$

$$\lim_{x\to1^+} f(x)=3 \qquad \lim_{x\to1^+} f'(x)=-1 \qquad \lim_{x\to1^+} f''(x)=0$$

$$\lim_{x\to1} f(x)=3 \qquad \lim_{x\to1} f'(x)=\text{does not exist} \qquad \lim_{x\to1} f''(x)=0$$

86. **(C)**

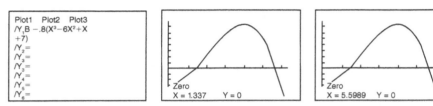

Since $v(t)$ represents the rate of change of the volume of water in the tub, $\int_{t=a}^{t=b} v(t)\,dt$ represents the changes of the volume in the tub between 2 times. The water is rising when rate of change is positive. The 40 gallons already in the tub doesn't enter into the solution.

87. **(A)**

By 2nd FTC, $\dfrac{d}{dx}\displaystyle\int_0^{\sin x} \dfrac{t}{1-t^2}\,dt = \left(\dfrac{\sin x}{1-\sin^2 x}\right)\cos x$

$= \left(\dfrac{\sin x}{\cos^2 x}\right)\cos x = \dfrac{\sin x}{\cos x} = \tan x$ \qquad OR

$\displaystyle\int_0^{\sin x} \dfrac{t}{1-t^2}\,dt = -\dfrac{1}{2}\ln\left(1-\sin^2 x\right) = -\dfrac{1}{2}\ln\left(\cos^2 x\right)$

$\dfrac{d}{dx}\left(-\dfrac{1}{2}\ln\left(\cos^2 x\right)\right) = \dfrac{-1}{2\cos^2 x}(-2\sin x \cos x) = \tan x$

88. **(A)**

$$\dfrac{2}{2}\left(2+k^2\right)+\dfrac{3}{2}\left(k^2+5\right)+\dfrac{2}{2}(5+8)=15k$$

$4 + 2k^2 + 3k^2 + 15 + 26 = 30k$

$5k^2 - 30k + 45 = 0$

$5(k^2 - 6k + 9) = 0 \Rightarrow (k-3)^2 = 0 \Rightarrow k = 3$

89. **(D)**

$$v(t) = \int \frac{4t-3}{e^{4t-3}}\, dt$$

$$v(1) - v(0) = \int_0^1 \frac{4t-3}{e^{4t-3}}\, dt \text{ so } v(0) = v(1) - \int_0^1 \frac{4t-3}{e^{4t-3}}\, dt$$

$$v(0) = 8.256 - (-10.227) = 18.483$$

90. **(B)**

$$\int_k^{\pi/2} (1-\cos x)\, dx = \frac{1}{2}$$

$$[x - \sin x]_k^{\pi/2} = \frac{1}{2}$$

$$\frac{\pi}{2} - 1 - (k - \sin k) = \frac{1}{2}$$

$$k - \sin k = \frac{\pi - 3}{2}$$

91. **(E)**

Since $f'(x)$ switches from negative to positive at $x = 0.440$, there is a local minimum at that location. However, since by looking at the graph of $f'(x)$, there is more negative area than positive area between $x = 0.440$ and $x = 2.5$, the absolute minimum must occur at $x = 2.5$.

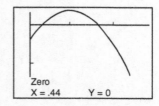

Zero
X = .44   Y = 0

92. **(E)**

The question asks for the average acceleration on $[0, 10]$.
Working backwards, III is correct by definition of the average value function.
Since $\int a(t)\, dt = v(t)$, I is correct.
And since $v = x'$, II is correct.

Detailed Explanations of Answers

Section II

Question 1

(a) $P'(-5) \approx \dfrac{P(0)-P(-15)}{0-(-15)} = \dfrac{175-420}{15} = -16.333$

The number of people in the mall is decreasing at a rate of approximately 16.333 people per minute 5 minutes before closing.

(b) $\displaystyle\int_{-25}^{30} P'(t)\ dt = P(30)-P(-25) = 25-750 = -725.$

725 people have left the mall between 25 minutes before closing and 30 minutes after closing.

$\dfrac{\displaystyle\int_{-25}^{30} P(t)\ dt}{30-(-25)} \approx \dfrac{10(420)+15(175)+10(90)+20(25)}{55} = \dfrac{8225}{55} = 149.545$

(c) This approximation is an underestimation because a right Riemann sum is used and the function P is strictly decreasing.

(d) $P(-5) = P(-15) + \displaystyle\int_{-15}^{-5} P'(t)\ dt$

$P(-5) = 420 - 191.429 = 228.571$

There were approximately 229 people in the mall 5 minutes before closing time.

Question 2

(a) $2\ln x = 8 - \dfrac{x^2}{2} \Rightarrow x = 3.343$

The graphs intersect at $(A, B) = (3.343, 2.414)$

$\text{Area} = \displaystyle\int_{1}^{A} 2\ln x\, dx + \int_{A}^{4}\left(8 - \dfrac{x^2}{2}\right) dx = 3.383 + 0.816 = 4.199$

OR $\quad y = 2\ln x \Rightarrow x = e^{y/2} \qquad y = 8 - \dfrac{x^2}{2} \Rightarrow x = \sqrt{16 - 2y}$

$$\text{Area} = \int_0^B \left(\sqrt{16 - 2y} - e^{y/2} \right) dy = 4.199$$

(b) $\quad \text{Volume} = \int_1^A (2\ln x)^2\, dx + \int_A^4 \left(8 - \dfrac{x^2}{2} \right)^2 dx$

(c) $\quad 2\int_0^k \left(\sqrt{16 - 2y} - e^{y/2} \right) dy = 4.199$

OR $\quad \int_0^k \left(\sqrt{16 - 2y} - e^{y/2} \right) dy = 2.1$

OR $\quad \int_0^k \left(\sqrt{16 - 2y} - e^{y/2} \right) dy = \int_k^{2.414} \left(\sqrt{16 - 2y} - e^{y/2} \right) dy$

Question 3

(a) $\quad g(6) = \displaystyle\int_{-2}^6 f(t)\, dt = \dfrac{-1}{2}\pi(1^2) + \dfrac{1}{2}\pi(2^2) + \dfrac{1}{2}(2)(8) = \dfrac{3\pi}{2} + 8.$

$\quad g(-6) = \displaystyle\int_{-2}^{-6} f(t)\, dt = \dfrac{1}{2}(-4)(-3) = 6.$

(b) $\quad g'(x) = \dfrac{d}{dx}\displaystyle\int_{-2}^x f(t)\, dt = f(x) \quad \Rightarrow \quad g'(5) = f(5) = 4.$

$\quad g''(x) = f'(x) \Rightarrow g''(5) = f'(5) = 4.$

(c) Critical points are where $g'(x) = f(x) = 0$ at $x = -6$, $x = -2$, $x = 0$, $x = 4$.

$$g'(x): \underline{+\!+\!+\!+\,0\,-\!-\!-\!-\!-\!-\!-\,0\,-\!-\!-\!-\,0\,+\!+\!+\!+\!+\!+\!+\!+\!+\!+\,0\,+\!+\!+\!+\!+}$$
$$\qquad\ \ -8 \qquad -6 \qquad\qquad -2 \qquad\ 0 \qquad\qquad\quad 4 \qquad\ 6$$

g' changes from positive to negative at $x = -6$ so g has a relative maximum at $x = -6$.

g' changes from negative to positive at $x = 0$ so g has a relative minimum at $x = 0$. No relative extrema at $x = -2$, $x = 4$.

(d) There are only two possibilities for the location of the absolute maximum: $x = -6$ and $x = 6$.

From part (a), $g(-6) = 6$ and $g(6) = \dfrac{3\pi}{2} + 8$, so the maximum value of g is

$\dfrac{3\pi}{2} + 8$.

(e) $g''(x) = f'(x):$ $------+++++++---++\;+\infty+++++---------+++++$
                        $-8 \qquad -5 \qquad\quad -2 \quad\; -1 \;\; -0 \qquad 2 \qquad\quad 4 \qquad\quad 6$

The graph of g has inflection points at each of $x = -5$, $x = -2$, $x = -1$, $x = 2$, and $x = 4$ because $g''(x) = f'(x)$ changes sign at these values.

Question 4

(a) $a(20) = v'(20) \approx \dfrac{v(25) - v(18)}{25 - 18} = \dfrac{-1+4}{25-18} = \dfrac{3}{7}\dfrac{m}{\sec^2}$

(b) $x(30) = x(0) + \displaystyle\int_0^{30} v(t)\, dt$

$x(30) \approx 6 + \dfrac{4}{2}(5+8) + \dfrac{8}{2}(8-8) + \dfrac{6}{2}(-8-4) + \dfrac{7}{2}(-4-1) + \dfrac{5}{2}(-1+4)$

$x(30) = 6 + 26 + 0 - 36 - \dfrac{35}{2} + \dfrac{15}{2} = -14$

(c) Avg speed $= \dfrac{\displaystyle\int_0^{30} |v(t)|\, dt}{30}$

$= \dfrac{\dfrac{4}{2}(5+8) + \dfrac{8}{2}(8+8) + \dfrac{6}{2}(8+4) + \dfrac{7}{2}(4+1) + \dfrac{5}{2}(1+4)}{30}$

$= \dfrac{26 + 64 + 36 + \dfrac{35}{2} + \dfrac{25}{2}}{30}$

$= \dfrac{156}{30} = \dfrac{26}{5}\dfrac{m}{\sec}$

(d) Since v is differentiable on $(4, 12)$, by the Mean Value Theorem there must be at least one time on $(4, 12)$ such that $v'(t) = a(t) = \dfrac{-8-8}{12-4} = -2\dfrac{m}{\sec^2}$.

(e) Since $a(t) > 0$ on $[0,4]$, $v(t) > 5$ on $[0, 4]$

So the minimum position of Newton is $x(4) = x(0) + \displaystyle\int_0^4 v(t)\, dt$

$x(4) > 6 + 4(5) = 26$.

Question 5

(a) $3y^2 \dfrac{dy}{dx} - 9\dfrac{dy}{dx} + 8x - 16 = 0$

$\dfrac{dy}{dx}\left(3y^2 - 9\right) = 16 - 8x$

$\dfrac{dy}{dx} = \dfrac{16 - 8x}{3y^2 - 9}$

(b) $\dfrac{16 - 8x}{3y^2 - 9} = 0$

$16 - 8x = 0$

$x = 2$

(c) $\dfrac{dy}{dx}\Big|_{(4,-2)} = \dfrac{16 - 8(4)}{3(-2)^2 - 9} = \dfrac{-16}{3}$

$y + 2 = \dfrac{-16}{3}(x - 4)$

$y = -2 - \dfrac{16}{3}(4.25 - 4) = -2 - \dfrac{4}{3} = \dfrac{-10}{3}$

(d) $\dfrac{d^2y}{dx^2} = \dfrac{-8\left(3y^2 - 9\right) - (16 - 8x)6y\dfrac{dy}{dx}}{\left(3y^2 - 9\right)^2}$

$\dfrac{d^2y}{dx^2}\Big|_{(4,-2)} = \dfrac{-8(3) - (-16)(-12)\left(\dfrac{-10}{3}\right)}{12^2} = \dfrac{-24 + 640}{144} > 0$

So the graph is concave up.

(e) Vertical tangents when $3y^2 - 9 = 0$ or $y = \pm\sqrt{3}$.
The curve crosses the y-axis when $x = 0$.
At $(0, \sqrt{3})$: $3\sqrt{3} - 9\sqrt{3} \approx 10$
At $(0, \sqrt{3})$: $-3\sqrt{3} + 9\sqrt{3} \approx 10$
Since neither point satisfies the original equation, there are no vertical tangents along the y-axis.

Question 6

(a) $\dfrac{dP}{dt}\Big|_{p=40} = \dfrac{1}{3}(100-40) = \dfrac{20\%}{\text{day}}$

$\dfrac{dP}{dt}\Big|_{p=60} = \dfrac{1}{3}(100-60) = \dfrac{13.333\%}{\text{day}}$

Since $\dfrac{dP}{dt}\Big|_{p=40} > \dfrac{dP}{dt}\Big|_{p=60}$, power is being restored faster when 40% of the city has power.

(b) $\dfrac{d^2P}{dt^2} = \dfrac{-1}{3}\dfrac{dP}{dt} = \dfrac{-1}{3}\cdot\dfrac{1}{3}(100-P) = -\dfrac{1}{9}(100-P)$

(c) Since $0 \le P \le 100$, $\dfrac{dP}{dt} > 0$, P is increasing.

Since $0 \le P \le 100$, $\dfrac{d^2P}{dt^2} < 0$, P is concave down.

$\lim\limits_{P\to100}\dfrac{dP}{dt} = 0$.

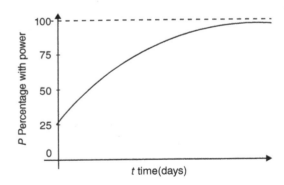

(d) $\dfrac{dP}{dt} = \dfrac{1}{3}(100-P)$

$\displaystyle\int\dfrac{1}{100-P} = \int\dfrac{1}{3}\,dt$

$-\ln|100-P| = \dfrac{1}{3}t + C$

$\ln|100-P| = -\dfrac{1}{3}t + C$

$100 - P = Ce^{-t/3}$

$P = 100 - Ce^{-t/3}$

$t = 0, P = 25 \Rightarrow 25 = 100 - Ce^0 \Rightarrow C = 75$

$P = 100 - 75e^{-t/3}$

AP Calculus BC
Practice Exam

Also available at the REA Study Center (*www.rea.com/studycenter*)

This practice exam is available at the REA Study Center. Although AP exams are administered in paper-and-pencil format, we recommend that you take the online version of the practice exam for the benefits of:

- Instant scoring
- Enforced time conditions
- Detailed score report of your strengths and weaknesses

AP Calculus BC Practice Exam
Section I
Part A

(Answer sheets appear in the back of the book.)

TIME: 55 minutes

Number of Questions—28

Directions: Solve each of the following problems, select the best answer choice, and fill in the corresponding oval on the answer sheet.

Calculators may NOT be used for this section of the exam.

Notes:

(1) Unless otherwise specified, the domain of a function f is assumed to be the set of real numbers x for which $f(x)$ is a real number.

(2) The inverse of a trigonometric function may be indicated using the inverse notation f^{-1} or with the prefix "arc" (e.g., $\sin^{-1} x = \arcsin x$).

1. If $y = \sqrt[3]{\cos^2 x}$, then $\dfrac{dy}{dx} =$

(A) $\dfrac{2}{3\sqrt[3]{\sin x}}$

(B) $\sqrt[3]{-2\sin x \cos x}$

(C) $\dfrac{2}{3\sqrt[3]{\cos x}}$

(D) $\dfrac{-2\sin x}{3\sqrt[3]{\cos x}}$

(E) $\dfrac{-3\sin x\sqrt{\cos x}}{2}$

2. The position of a particle moving in the xy-plane is given by the parametric equations $x(t) = t^2 - 4t$ and $y(t) = 16t - 2t^2$. At which of the following points is the particle at rest?

(A) $(2, 4)$

(B) $(-4, 32)$

(C) $(4, 8)$

(D) A different point than the points above

(E) It is never at rest

3. $\displaystyle\int_1^8 \frac{dx}{x^{2/3}\left(1+x^{1/3}\right)}$ is equivalent to which of the following?

(A) $\dfrac{1}{3}\displaystyle\int_1^2 \dfrac{du}{1+u}$

(B) $\displaystyle\int_1^2 \dfrac{3du}{1+u}$

(C) $\displaystyle\int_2^3 \dfrac{3du}{1+u}$

(D) $\displaystyle\int_2^3 \dfrac{du}{3u}$

(E) $\displaystyle\int_2^3 \dfrac{3du}{u}$

4. Derek likes to collect apps on his smartphone. He downloads apps at the rate of $A(t)$ where $A(t)$ is measured in apps per day and t is measured in days. Selected values of $A(t)$ are given in the table below. Using a trapezoidal rule with four intervals and data from the table, what is the approximation for the average number of apps he downloads per week?

t (days)	0	3	7	11	14
$A(t)$ (apps/day)	4	2	1	3	5

(A) 2.5

(B) 2.625

(C) 17.5

(D) 18.375

(E) 35

5. Find the length of the curve $y = |x^2 - 1|$ from $x = -2$ to $x = 3$.

(A) $\displaystyle\int_{-2}^{3} \sqrt{1+4x^2}\,dx$

(B) $\displaystyle\int_{-2}^{3} \sqrt{1+2x}\,dx$

(C) $\displaystyle\int_{-2}^{3} \left(1+4x^2\right)dx$

(D) $\displaystyle\int_{-2}^{0} \sqrt{1-2x}\,dx + \int_{0}^{3} \sqrt{1+2x}\,dx$

(E) $\displaystyle\int_{-2}^{3} \left(1+2x\right)^2 dx$

6. The Maclaurin series for the function f is given by $f(x) = \displaystyle\sum_{n=1}^{\infty} \left(\dfrac{-2x}{5}\right)^n$. What is the value of $f(2)$?

(A) 5

(B) 4

(C) $\dfrac{5}{9}$

(D) $\dfrac{1}{5}$

(E) $-\dfrac{4}{9}$

7. If $\cos^{-1} x = e^y$, which of the following is an expression for $\dfrac{dy}{dx}$?

I. $\dfrac{-1}{e^y \sqrt{1-x^2}}$

II. $\dfrac{-1}{\cos^{-1} x \sqrt{1-x^2}}$

III. $\dfrac{-\csc\left(e^y\right)}{e^y}$

(A) I only

(B) II only

(C) III only

(D) I and II only

(E) I, II, and III

8. The three regions A, B, and C in the figure below are bounded by the graph of the function f and the x-axis. If the areas of A, B, and C are 8, 3, and 1, respectively, what is the value of $\int_{-4}^{1} [f(x) - 2x]\, dx$?

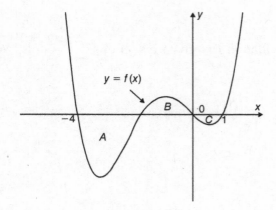

(A) 4

(B) 9

(C) 22

(D) 27

(E) −21

9. What is the radius of convergence of $\sum_{n=0}^{\infty} \dfrac{(4x)^{2n}}{9^n}$?

(A) 1

(B) $\dfrac{3}{4}$

(C) $\dfrac{3}{2}$

(D) $\dfrac{2}{3}$

(E) $\dfrac{4}{9}$

10. Mail flow rate $R'(t)$ is defined as the rate in which letters pass through a sorter at a mail facility, measured in letters per minute. Which of the following gives the average rate of change of the flow rate from $t = 10$ to $t = 20$ minutes?

I. $\dfrac{R'(20) - R'(10)}{10}$

II. $\dfrac{\displaystyle\int_{10}^{20} R'(t)\,dt}{10}$

III. $\dfrac{\displaystyle\int_{10}^{20} R''(t)\,dt}{10}$

(A) I only

(B) II only

(C) III only

(D) I and II only

(E) I and III only

11. The graph of f', the derivative of f consists of two line segments and a semicircle, as shown in the figure below. If $f(-2) = -1$, then $f(5) =$

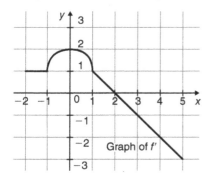

Graph of f'

(A) $\dfrac{\pi}{2}$

(B) $\dfrac{\pi}{2} + 1$

(C) $\dfrac{\pi}{2} - 1$

(D) $\dfrac{\pi}{2} + 2$

(E) $\dfrac{\pi}{2} - 2$

12. Find $\displaystyle\int_1^2 \frac{x-1}{x^2-7x+12}dx$

   (A) $\ln\left(\dfrac{9}{10}\right)$

   (B) $\ln\left(\dfrac{10}{9}\right)$

   (C) $\ln\left(\dfrac{32}{27}\right)$

   (D) $\ln\left(\dfrac{2}{27}\right)$

   (E) Divergent

13. The graph of the function $y = f(x)$ consists of 2 line segments as shown below. Which of the following statements about f is *false*?

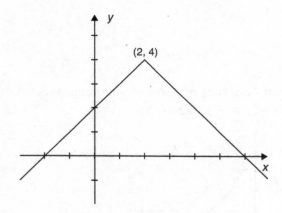

   (A) $\displaystyle\lim_{x\to 2}[f(x)-f(2)]=0$

   (B) $\displaystyle\lim_{x\to 2}\frac{f(x)-f(2)}{x-2}=0$

   (C) $\displaystyle\lim_{x\to 2}\frac{f(x+h)-f(x-h)}{2h}=0$

   (D) $\displaystyle\lim_{x\to 0}\frac{f(x)-f(2)}{x-2}=1$

   (E) $f'(x)$ is not continuous

14. If you place an ice cube in a pool of water that is 72°F, the water at the edge of the ice cube will be 32°F. The further from the ice cube, the warmer the water. If t is the temperature of the water around the ice cube and x is the distance from the edge of the ice cube, then $\dfrac{dt}{dx} = 72 - t$. Write an expression that predicts the temperature of the water at distance x from the edge of the ice cube.

(A) $t = 72 + 32e^{-x}$

(B) $t = 72 - 32e^x$

(C) $t = 32e^{-x}$

(D) $t = 72 - 40e^{-x}$

(E) $t = 72 + 40e^x$

15. Let $y = f(x)$ be the solution to the differential equation $\dfrac{dy}{dx} = 2x - y^2$ with initial condition $f(-1) = -2$. What is the approximation for $f(0)$ obtained by using Euler's method with two steps starting at $x = -1$?

(A) -31

(B) -18

(C) -2

(D) $\dfrac{-3}{2}$

(E) 7

16. For $x > 0$, the power series $1 - \dfrac{x^4}{3!} + \dfrac{x^8}{5!} - \dfrac{x^{12}}{7!} + \ldots + (-1)^n \dfrac{x^{4n}}{(2n+1)!} + \ldots$ converges to which of the following?

(A) $\cos x^2$

(B) e^{-4x}

(C) $\sin x^4$

(D) $\dfrac{\cos x^2 - 1}{x}$

(E) $\dfrac{\sin x^2}{x^2}$

17. When a popular band comes to a town to perform, people line up to purchase tickets. The number of people waiting to purchase tickets is modeled by a twice-differentiable function N of t, where t is measured in minutes. For $0 < t < 10$, $N''(t) > 0$. The table below gives selected values of the rate of change $N'(t)$ over the time interval $0 \le t \le 10$. The number of people in line at $t = 5$ is 22. By using the tangent line approximation at $t = 5$, which of the following is a possible value for the people in line at $t = 5.5$ minutes?

t (minutes)	0	2	5	7	10
$N'(t)$ (people per minute)	3	8	12	26	40

I. 28

II. 30

III. 36

(A) I only

(B) II only

(C) III only

(D) I and II only

(E) I, II, and III

18. The graph of $f(x)$ is a straight line passing through the origin as shown in the figure below.

What is $\displaystyle\lim_{x \to 4} \frac{\displaystyle\int_{-4}^{x} f(t)\,dt}{f(x)+8}$?

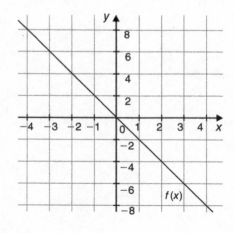

(A) −4

(B) 4

(C) −1

(D) 1

(E) Does not exist

19. A water tank has the shape and dimensions as shown in the figure below. At time $t = 0$, the tank is empty. Then it is filled with water at the rate of $1\dfrac{\text{ft}^3}{\text{min}}$. If h represents the height of the water at time t, which of the following graphs represents $\dfrac{dh}{dt}$ over time?

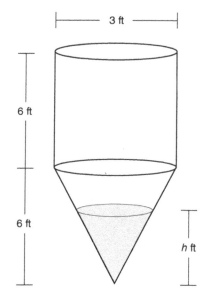

(A)

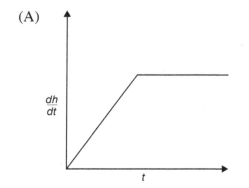

(B)

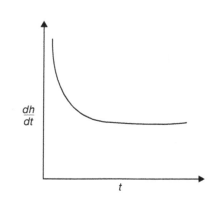

(C)
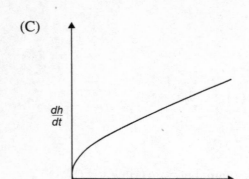

(D)
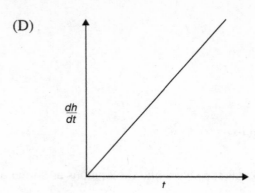

(E)
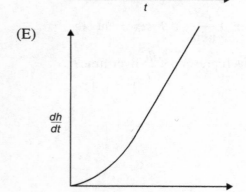

20. Which of the following series converge?

I. $\displaystyle\sum_{n=1}^{\infty} \frac{3^n}{(n-1)!}$

II. $\displaystyle\sum_{n=1}^{\infty} \frac{n!}{n^3}$

III. $\displaystyle\sum_{n=0}^{\infty} \frac{n+3}{(n+1)(n+2)}$

(A) I only

(B) II only

(C) III only

(D) I and II only

(E) I and III only

21. $\int\limits_{-\infty}^{2} x^2 e^{x^3} \, dx =$

    (A) $\dfrac{e}{3}$

    (B) $\dfrac{e^4}{3}$

    (C) $\dfrac{e^8}{3}$

    (D) $\dfrac{e-1}{3}$

    (E) $\dfrac{e^8 - 1}{3}$

22. The points $(-3, 4)$ and $(2, -3)$ are on the graph of a function $y = f(x)$ that satisfies the differential equation $\dfrac{dy}{dx} = x^2 + xy + 2$. Which of the following must be true?

    (A) $(-3, 4)$ is a local minimum of f

    (B) $(-3, 4)$ is an inflection point of the graph of f

    (C) $(2, -3)$ is a local minimum of f

    (D) $(2, -3)$ is a local maximum of f

    (E) $(2, -3)$ is a point of inflection of the graph of f

23. There are 25,000 workers in the Pentagon. A new smartphone becomes available and the rate of change $\dfrac{dP}{dt}$ of the number of people in the Pentagon who purchase this new smartphone is modeled by a logistic differential equation. On January 1, the number of people who have the smartphone is 5,000 and is increasing at the rate of 2,500 people per month. Which of the following differential equations models the situation?

    (A) $\dfrac{dP}{dt} = \dfrac{1}{2} P(2500 - P)$

    (B) $\dfrac{dP}{dt} = \dfrac{1}{2}(25000 - P)$

    (C) $\dfrac{dP}{dt} = \dfrac{1}{40000}(25000 - P) + 5000$

    (D) $\dfrac{dP}{dt} = 40000 \, P(25000 - P)$

    (E) $\dfrac{dP}{dt} = \dfrac{1}{40000} P(25000 - P)$

24. Let f be a function having derivatives for all orders of real numbers. The first three derivatives of f at $x = 1$ are given in the table below. Use the third-degree Taylor polynomial at $x = 1$ to approximate $f\left(\dfrac{1}{2}\right)$.

x	$f(x)$	$f'(x)$	$f''(x)$	$f'''(x)$
1	3	4	-4	-48

(A) -2

(B) -3

(C) $\dfrac{3}{2}$

(D) $\dfrac{7}{2}$

(E) 6

25. The value of Frank's stock over a period of a week is given by the function S. At $t = 0$, the value of S is \$10,000. The graph of $\dfrac{dS}{dt}$ over the period of a week is shown in the figure below with the areas between the curve and the horizontal axis indicated in dollars. What is the difference between the maximum and minimum value of Frank's stock over the period of the week?

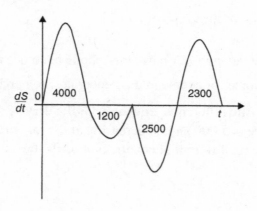

(A) \$0

(B) \$2,500

(C) \$3,700

(D) \$4,000

(E) \$10,000

26. Let f be a function such that $\int \dfrac{x^2}{1+x^2}\,dx = x^2 f(x) - 2\int x f(x)\,dx$. Which of the following could be $f(x)$?

    (A) $\ln x$

    (B) $\ln(1 + x^2)$

    (C) $\ln(1 + x)$

    (D) $\tan^{-1} x$

    (E) $\sin^{-1} x$

27. Find the equation of the tangent line to the polar curve $r = e^\theta$ at $\theta = \dfrac{\pi}{2}$.

    (A) $y = e^{\pi/2}$

    (B) $y = e^{\pi/2} - x$

    (C) $y = e^{\pi/2} + x$

    (D) $y = -x - 2e^{\pi/2}$

    (E) $y = x + 2e^{\pi/2}$

28. A small lighthouse has a base section and a light section as shown in the figure below. The 10-meter-high base is formed by rotating the portion of the line $y = 15 - 15x$ above the x-axis about the y-axis. The light section has height 2. Which of the following represents the volume of the lighthouse?

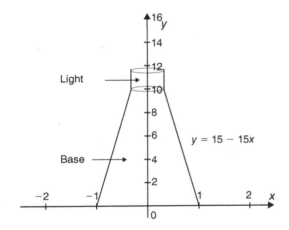

(A) $\pi\int_0^{10}\left(1-\dfrac{y}{15}\right)^2 dy+\dfrac{2\pi}{9}$

(B) $\pi\int_0^{12}\left(1-\dfrac{y}{15}\right)^2 dy$

(C) $\pi\int_0^1 (15-15x)^2 dx+\dfrac{2\pi}{3}$

(D) $\pi\int_0^1 (15-15x)^2 dx+\dfrac{\pi}{9}$

(E) $2\pi\int_0^{12}\left(1-\dfrac{y}{15}\right)^2 dy$

STOP

This is the end of Section I, Part A.

If time still remains, you may check your work only in this section.

Do not begin Section I, Part B until instructed to do so.

Section I

Part B

TIME: 50 minutes

Number of Questions—17

Directions: Solve each of the following problems, select the best answer choice, and fill in the corresponding oval on the answer sheet.

A graphing calculator is required for some questions on this part of the exam.

Notes:

(1) The exact numerical value of the correct answer does not always appear among the choices given. When this happens, select from among the choices the number that best approximates the exact numerical value.

(2) Unless otherwise specified, the domain of a function f is assumed to be the set of real numbers x for which $f(x)$ is a real number.

(3) The inverse of a trigonometric function may be indicated using the inverse notation f^{-1} or with the prefix "arc" (e.g., $\sin^{-1} x = \arcsin x$).

76. The figure below shows the graph of a function f. Which of the following has the smallest value?

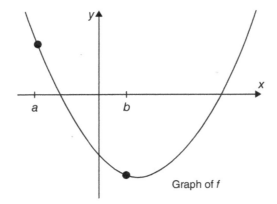

Graph of f

(A) $f(a)$

(B) $f'(a)$

(C) $\lim\limits_{x \to b} \dfrac{f(x) - f(b)}{x - b}$

(D) $\dfrac{f(b) - f(a)}{b - a}$

(E) $f''(b)$

77. The third derivative of the function f is continuous on the interval $(-2, 0)$. Values for f and its first three derivatives at $x = -1$ are given in the table below. Find
$$\lim\limits_{x \to -1} \dfrac{f(x)}{\cos(\pi x) + 1}.$$

x	$f(x)$	$f'(x)$	$f''(x)$	$f'''(x)$
-1	0	0	-3	8

(A) 0

(B) -3

(C) 3

(D) $\dfrac{-3}{\pi^2}$

(E) $\dfrac{3}{\pi^2}$

78. When people walk into a supermarket, there is a large stack of coupons for a free item. The rate at which this pile decreases is modeled by $r(t) = 100\left(\sqrt{t} + 2\sin t - 4\right)$, where t is the number of hours after the market opens, $0 \le t < 6$. At how many hours after the store opens is the pile of coupons decreasing the most rapidly?

(A) 0

(B) 2.292

(C) 3.164

(D) 4.596

(E) 6

79. The graph in the figure below is $f(x) = 2.01 - \sqrt{\sin^2 x + 0.0001}$. Which of the following is true?

    I. $\lim\limits_{x \to 0} f(x) = 2$

    II. f is continuous at $x = 0$

    III. f is differentiable at $x = 0$

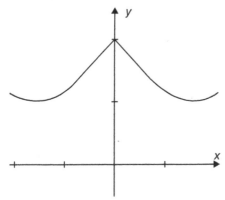

    (A) I only

    (B) II only

    (C) I and II only

    (D) I, II, and III

    (E) None are true

80. The table below shows several Riemann sum approximations to $\int_0^1 (e^x - e^{-x})\, dx$ using right-hand endpoints of n subintervals of equal length of the interval $[0, 1]$. Which of the following statements best describes the limit of the Riemann sums as n approaches infinity?

n	$\sum\limits_{k=1}^{n} \left(e^{k/n} - e^{-k/n}\right)\left(\dfrac{1}{n}\right)$
5	1.325
10	1.204
15	1.165
20	1.145
25	1.133

(A) The limit of the Riemann sums is a finite number less than 1.

(B) The limit of the Riemann sums is a finite number greater than 1.

(C) The limit of the Riemann sums does not exist as $\left(e^{k/n} - e^{-k/n}\right)\left(\dfrac{1}{n}\right)$ does not approach zero.

(D) The limit of the Riemann sums does not exist because it is a sum of infinitely many positive numbers.

(E) The limit of the Riemann sums does not exist because $\displaystyle\int_{0}^{1}\left(e^{x} - e^{-x}\right) dx$ is an improper integral.

81. A power series is used to approximate $\displaystyle\int_{0}^{1} x \sin x \, dx$ with a maximum error of 0.001. What is the minimum number of terms needed to obtain this approximation?

(A) One

(B) Two

(C) Three

(D) Four

(E) Five

82. The graph below shows $f(x) = \cos x(1 - \sin x) + 1$ on the interval $[0, \pi]$. There are two values of c that satisfy the Mean Value Theorem for f between points $P(0, 2)$ and $Q(\pi, 0)$. Find the area between the graph of f and the x-axis between these two values of c.

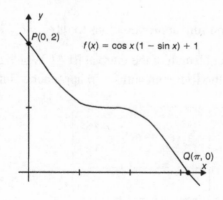

(A) 1.071

(B) 1.463

(C) 2.071

(D) 2.648

(E) 3.142

83. The Taylor polynomial of degree 75 for the function f about $x = 0$ is given by
$P(x) = \dfrac{x}{2!} - \dfrac{x^3}{6!} + \dfrac{x^5}{10!} - \dfrac{x^7}{14!} + \ldots + (-1)^{n+1} \dfrac{x^{2n+1}}{(4n+2)!} + \ldots - \dfrac{x^{75}}{150!}$. What is the value of $f^{(25)}(0)$?

(A) $\dfrac{25!}{50!}$

(B) $-\dfrac{25!}{50!}$

(C) $\dfrac{50!}{25!}$

(D) $-\dfrac{50!}{25!}$

(E) $\dfrac{1}{2}$

84. For what values of p does the infinite series $\displaystyle\sum_{n=1}^{\infty}(-1)^{n+1}\dfrac{n}{n^{3p-4}+1}$ converge?

(A) $p > \dfrac{5}{3}$

(B) $p > \dfrac{1}{2}$

(C) $p \geq \dfrac{1}{2}$

(D) $p > 2$

(E) $p \geq 2$

85. Inside a square, there is an inscribed circle of radius r. Outside the square there is a circumscribed circle as shown in the figure below. At a certain instant the rate of increase of the area of the inscribed circle is equal to the rate of increase of the circumference of the circumscribed circle. What is the area of the square at that instant?

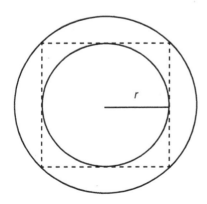

(A) 1

(B) $\sqrt{2}$

(C) 2

(D) 4

(E) 8

86. Let $f(x) = \sum_{n=1}^{\infty} \frac{x^{n+1}}{n(n+1)}$. Find the interval of convergence of $\int f(x)\, dx$.

(A) $(-1, 1)$

(B) $[-1, 1)$

(C) $(-1, 1]$

(D) $[-1, 1]$

(E) The series is not convergent.

87. Let R be the region in the first quadrant bounded by the graphs of $y = e^{-x/2}, y = \frac{x}{2}$, and the y-axis as shown in the figure below. The graphs intersect at the point $(1.134, 0.567)$. R is the base of a solid for which each cross section perpendicular to the x-axis is an isosceles right triangle with the distance L being one of the legs of the right triangle. What is the volume of the solid?

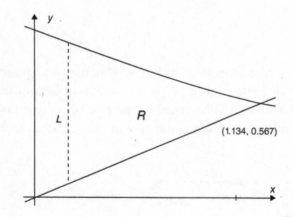

(A) 0.089

(B) 0.178

(C) 0.355

(D) 0.588

(E) 1.116

88. The function f has derivatives of all orders for all real numbers and $f^{(4)}(x) = e^{\sin x + \cos x}$. If the third-degree Taylor polynomial for f about $x = 0$ is used to approximate f on the interval $[0, 1]$, what is the Lagrange error bound for the maximum error on $[0, 1]$?

   (A) 0.033

   (B) 0.166

   (C) 0.171

   (D) 1.028

   (E) 4.113

89. The graph below shows the polar curve $r = 2\theta + 3\cos\theta$ for $0 \le \theta \le \pi$. What is the difference between the first quadrant area and the second quadrant area?

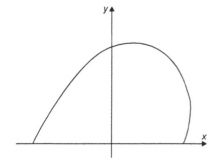

   (A) 0.533

   (B) 3.346

   (C) 6.693

   (D) 30.277

   (E) 60.555

90. A chipmunk is running along a horizontal roof. The velocity v of the chipmunk at time t, $0 \le t \le 6$ is given by the function whose graph is below. At what value of t does the chipmunk change direction?

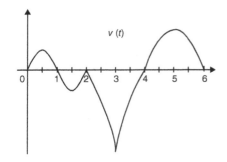

(A) $t = 0.5, 1.5,$ and 5 only

(B) $t = 0.5, 1.5, 3,$ and 5 only

(C) $t = 1$ and 4 only

(D) $t = 1, 2,$ and 4 only

(E) $t = 1, 2, 3,$ and 4 only

91. The function f has a continuous derivative. The table below gives values of f and its derivative for $x = -1$ and $x = 6$. If $\int\limits_{-1}^{6} f(x)\, dx = 15$, what is the value of $\int\limits_{-1}^{6} x \cdot f'(x)\, dx$?

x	$f(x)$	$f'(x)$
-1	3	2
6	-4	5

(A) -42

(B) -36

(C) -26

(D) -14

(E) 6

92. A particle, initially at rest, moves along a curve in the xy-plane with acceleration vector $\langle 2\pi \sin \pi t, e^t \rangle$. Find the speed of the particle at $t = 1$.

(A) 2.718

(B) 3.375

(C) 3.718

(D) 4.353

(E) 6.846

STOP

This is the end of Section I, Part B.

If time still remains, you may check your work only in this section.

Do not begin Section II, Part A until instructed to do so.

Section II

Part A

Free-Response Questions

TIME: 30 minutes

2 problems

Directions: Show all your work in your exam booklet. Grading is based on the methods used to solve the problems as well as the accuracy of your final answers.

A graphing calculator is required for this section.

Notes:

(1)   Unless otherwise specified, your final answers should be accurate to three decimal places.

(2)   Unless otherwise specified, the domain of a function f is assumed to be the set of all real numbers x for which $f(x)$ is a real number.

Question 1

A 4-hour carnival is serving free steaks. The rate that steaks are being grilled is given by the non-continuous function $C(t) = \begin{cases} 364 \text{ for } t \leq 2 \\ 144 \text{ for } t > 2 \end{cases}$. The rate that the steaks are eaten is given by $E(t) = 400 - 25t^2$ as shown in the figure below. Both rates are measured in steaks per hour and t is measured in hours. When the steaks are done, they are put in warming trays. There are 150 steaks that are already in the warming trays when the carnival opens.

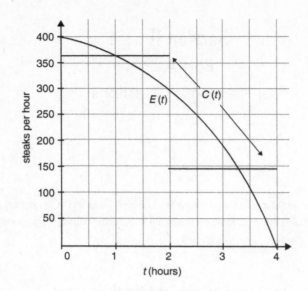

(a) What is the rate of change of $E(t)$ at $t = 1.5$? Explain your answer using proper units.

(b) What is the average number of steaks eaten each hour (nearest steak)?

(c) For $0 \leq t \leq 4$, at what time t is the number of steaks that are either on the grill or in the warming trays a minimum?

(d) The carnival stays open after its announced 4 hours. During that time, no more steaks are grilled, but the remaining steaks in the warming tray are eaten at the rate of $50\sqrt{t-4}$, $t \geq 4$. Write, but do not solve, an equation that determines the amount of time k needed for people to eat the remaining steaks.

Question 2

For $t \geq 0$, a particle is moving along a curve so that its position at time is $(x(t), y(t))$. At time $t = 1$, the particle is at position $(-\pi, -1)$. It is known that

$$\frac{dx}{dt} = t - \cos^2 t \quad \text{and} \quad \frac{dy}{dt} = \frac{1 - 4\ln(t+1)}{t+1}.$$

(a) Is the horizontal movement of the particle to the left or right at time $t = 5$? Explain your answer. Find the slope of the path of the particle at time $t = 5$.

(b) Find the position of the particle at time $t = 5$.

(c) Find the speed and acceleration vector of the particle at time $t = 5$.

(d) Find the distance traveled by the particle from time $t = 0$ to $t = 5$.

Section II
Part B

TIME: 60 minutes

4 problems

Directions: Show all your work in your exam booklet. Grading is based on the methods used to solve the problems as well as the accuracy of your final answers.

No calculator is allowed for this section.

Question 3

The figure below shows the graphs of the line $y = \dfrac{2x}{3}$ and curve C given by $x = \dfrac{\sqrt{y^2 + 32}}{2}$

in the first quadrant. Let R be the region bounded by the two graphs and the x-axis. The line and the curve intersect at point P.

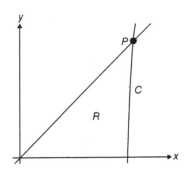

(a) Find the coordinates of point P and the value of $\dfrac{dx}{dy}$ for curve C at point P.

(b) The base of a solid is region R. Each cross section of the solid perpendicular to the y-axis is a semicircle. Write, but do not evaluate, an expression that gives the volume of the solid.

(c) Show that curve C can be written as the polar equation $r^2 = \dfrac{32}{4\cos^2\theta - \sin^2\theta}$.

(d) Use the polar equation in part (c) to set up an integral expression with respect to the polar angle θ that represents the area of R.

Question 4

Consider the differential equation $\dfrac{dy}{dx} = y - x - 1$.

(a) On the axes below, sketch a slope field for the given differential equation at the twelve points indicated and sketch the solution curve passing through (1, 2).

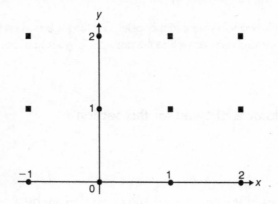

(b) The solution curve that passes through the point (1, 2) has a relative maximum at $x = e$. What is the y-coordinate of this relative maximum?

(c) Let $y = f(x)$ be the particular solution to the given differential equation with the initial condition $f(1) = 2$. Use Euler's method starting at $x = 1$ with two steps of equal size to approximate $f(0.5)$. Show the work that leads to your answer.

(d) On the graph below, shade the area that represents all points in which the graph of $f(x)$ is increasing and concave down.

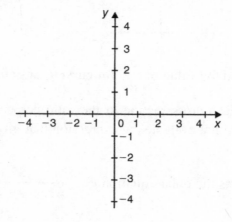

(e) If $f(1) = 2$, find the second-degree Taylor polynomial for f about $x = 1$.

Question 5

A particle is moving along the x-axis with velocity $v(t) = t^2 e^{-t}$ on the interval $[0, \infty)$. At $t = 0$, the particle is at position $x = 1$. Assume $t \geq 0$.

(a) When does the particle's velocity reach its maximum value? Justify your answer.

(b) Show that as t gets infinitely large, the particle is always slowing down.

(c) Find the position of the particle at $t = 2$.

(d) How far does the particle travel after $t = 2$? Explain your reasoning.

Question 6

The function f is defined by the power series

$$f(x) = 1 - \left(x + \frac{1}{2}\right) + \left(x + \frac{1}{2}\right)^2 - \left(x + \frac{1}{2}\right)^3 + \ldots + \left(x + \frac{1}{2}\right)^n + \ldots = \sum_{n=0}^{\infty} (-1)^n \left(x + \frac{1}{2}\right)^n$$

for all real numbers x for which the series converges.

(a) Find the interval of convergence of the power series for f. Justify your answer.

(b) This power series is the Taylor series for f about $x = -\frac{1}{2}$. Find the sum of the series for f and explain your reasoning.

(c) Let g be the function defined by $g(x) = f\left(x^2 - \frac{1}{2}\right)$. Find the first three non-zero terms of the Taylor series for g about $x = 0$, the general term, and find the value of $g\left(\sqrt{\frac{1}{3}}\right)$.

Answer Key

Section I

Part A

1. (D)	15. (B)		
2. (E)	16. (E)		
3. (E)	17. (B)		
4. (C)	18. (B)		
5. (A)	19. (B)		
6. (E)	20. (A)		
7. (E)	21. (C)		
8. (B)	22. (C)		
9. (B)	23. (E)		
10. (E)	24. (C)		
11. (E)	25. (D)		
12. (C)	26. (D)		
13. (B)	27. (B)		
14. (D)	28. (A)		

Part B

76. (B)	90. (C)
77. (D)	91. (B)
78. (A)	92. (D)
79. (D)	
80. (B)	
81. (C)	
82. (B)	
83. (A)	
84. (E)	
85. (E)	
86. (D)	
87. (B)	
88. (C)	
89. (B)	

Detailed Explanations of Answers

Section I, Part A

1. **(D)**

$$y = (\cos x)^{2/3}$$

$$\frac{dy}{dx} = \frac{2}{3}(\cos x)^{-1/3}(-\sin x) = \frac{-2\sin x}{3\sqrt[3]{\cos x}}$$

2. **(E)**

$$x'(t) = 2t - 4 \text{ and } y'(t) = 16 - 4t$$

$$2t - 4 = 0 \Rightarrow t = 2 \quad 16 - 4t = 0 \Rightarrow t = 4$$

Since $x'(t)$ and $y'(t)$ are never simultaneously zero, the particle is never at rest.

3. **(E)**

$$u = 1 + x^{1/3}, \quad du = \frac{1}{3x^{2/3}}dx$$

$$x = 1, u = 2 \quad x = 8, u = 3$$

$$\int_1^8 \frac{dx}{x^{2/3}\left(1 + x^{1/3}\right)} = 3\int_1^8 \frac{dx}{3x^{2/3}\left(1 + x^{1/3}\right)} = 3\int_2^3 \frac{du}{u} = \int_2^3 \frac{3du}{u}$$

4. **(C)**

$$\frac{\int_0^{14} A(t)\,dt}{14} = \frac{\frac{3}{2}(4+2) + \frac{4}{2}(2+1) + \frac{4}{2}(1+3) + \frac{3}{2}(3+5)}{14}$$

$$= \frac{9+6+8+12}{14} = \frac{35}{14}\frac{\text{apps}}{\text{day}} = \frac{35}{2}\frac{\text{apps}}{\text{week}}$$

5. **(A)**

It should be obvious by the graph below that the arc length of $|x^2 - 1|$ is the same as the arc length of $y = x^2 - 1$. $\dfrac{dy}{dx} = 2x$, so $\left(\dfrac{dy}{dx}\right)^2 = 4x^2$. $L = \int_{-2}^{3} \sqrt{1 + 4x^2}\, dx$.

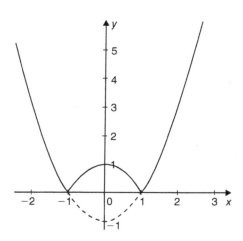

6. **(E)**

$f(2) = \displaystyle\sum_{n=1}^{\infty} \left(\dfrac{-4}{5}\right)^n$ is a convergent geometric series.

$$\sum_{n=0}^{\infty} \left(\dfrac{-4}{5}\right)^n = \dfrac{1}{1 - \left(\dfrac{-4}{5}\right)} = \dfrac{5}{5 + 4} = \dfrac{5}{9}.$$

So $\displaystyle\sum_{n=1}^{\infty} \left(\dfrac{-4}{5}\right)^n = \dfrac{5}{9} - \left(\dfrac{-4}{5}\right)^0 = \dfrac{5}{9} - 1 = \dfrac{-4}{9}.$

7. **(E)**

$$\dfrac{-1}{\sqrt{1 - x^2}} = e^y \dfrac{dy}{dx}$$

$$\dfrac{dy}{dx} = \dfrac{-1}{e^y \sqrt{1 - x^2}} = \dfrac{-1}{\cos^{-1} x \sqrt{1 - x^2}}$$

$$\cos^{-1} x = e^y \Rightarrow x = \cos(e^y)$$

$$1 = -\sin(e^y) \cdot (e^y) \dfrac{dy}{dx}$$

$$\dfrac{dy}{dx} = \dfrac{1}{-\sin(e^y) \cdot (e^y)} = \dfrac{-\csc(e^y)}{e^y}$$

8. **(B)**

$$\int_{-4}^{1} [f(x)-2x] \, dx = \int_{-4}^{1} f(x) \, dx - \int_{-4}^{1} 2x \, dx$$

$$-8+3-1-[x^2]_{-4}^{1} = -6-[1-16] = -6+15 = 9$$

9. **(B)**

Ratio test: $\lim_{n\to\infty} \left[\dfrac{(4x)^{2n+2}}{9^{n+1}} \cdot \dfrac{9^n}{(4x)^{2n}} \right] = \lim_{n\to\infty} \left[\dfrac{16x^2}{9} \right] < 1$

$$\pm\frac{4x}{3} < 1 \Rightarrow \pm 4x < 3$$

$$\frac{-3}{4} < x < \frac{3}{4} \text{ so the radius of convergence is } \frac{3}{4}.$$

No need to check the endpoints because the interval of convergence was not asked for.

10. **(E)**

The definition of the average rate of change of f over $[a,b]$ is $\dfrac{f(b)-f(a)}{b-a}$.

Substitute flow rate R' for f and I is correct.

Since $\int R''(t)\,dt = R'$, III is correct as well.

11. **(E)**

$$f(x) = \int f'(x) \, dx$$

$$f(5)-f(-2) = \int_{-2}^{5} f'(x) \, dx$$

$$f(5)-(-1) = 3 + \frac{\pi}{2} + \frac{1}{2} - \frac{9}{2}$$

$$f(5) = \frac{\pi}{2} - 2$$

12. **(C)**

$$\int_1^2 \frac{x-1}{x^2-7x+12}\,dx = \int_1^2 \frac{x-1}{(x-4)(x-3)}\,dx$$

$$\int_1^2 \left(\frac{3}{x-4} - \frac{2}{x-3}\right)dx$$

$$\left[3\ln|x-4| - 2\ln|x-3|\right]_1^2$$

$$3\ln 2 - 2\ln 1 - (3\ln 3 - 2\ln 2)$$

$$5\ln 2 - 3\ln 3$$

$$\ln 2^5 - \ln 3^3 = \ln 32 - \ln 27 = \ln\left(\frac{32}{27}\right)$$

13. **(B)**

(A) Plugging in 2, we get $f(2) - f(2) = 0$.

(B) This states that the derivative at $x = 2$ exists. Since this is a cusp point, it is not true.

(C) This takes a point h units to the right and left of 2 and subtracts them. The slope of this line is always zero.

(D) Plugging in, $\dfrac{f(0) - f(2)}{0 - 2} = \dfrac{2 - 4}{-2} = 1$

(E) To the left of (2, 4), the derivative is 2 and to the right it is -2. So the derivative is not continuous.

14. **(D)**

In this problem, you have to solve for t in terms of x, not the usual x in terms of t.

$$\int \frac{dt}{t-72} = \int -1\ dx$$

$$\ln|t-72| = -x + C$$

$$t - 72 = Ce^{-x} \Rightarrow t = 72 + Ce^{-x}$$

$$x = 0 : t = 32 \Rightarrow 32 = 72 + C \Rightarrow C = -40$$

$$t = 72 - 40e^{-x}$$

15. **(B)**

x	-1	-0.5	0
$y_{new} = y_{old} + \dfrac{dy}{dx}\Delta x$	-2	$-2 - 6(0.5) = -5$	$-5 - 26(0.5) = -18$
$\dfrac{dy}{dx}$	-6	-26	

16. **(E)**

It is best to do this problem by trial and error using these power series:

$$\sin x = x - \frac{x^3}{3!} + \frac{x^5}{5!} + \ldots \quad \cos x = 1 - \frac{x^2}{2!} + \frac{x^4}{4!} + \ldots \quad e^x = 1 + x + \frac{x^2}{2!} + \frac{x^3}{3!} + \ldots$$

(A) $\cos x^2 = 1 - \frac{x^4}{2!} + \frac{x^8}{4!} + \ldots$

(B) $e^{-4x} = 1 - 4x + \frac{16x^2}{2!} - \frac{64x^3}{3!} + \ldots$

(C) $\sin x^4 = x^4 - \frac{x^{12}}{3!} + \frac{x^{20}}{5!} + \ldots$

(D) $\frac{\cos x^2 - 1}{x} = -\frac{x^3}{2!} + \frac{x^7}{4!} + \ldots$

(E) $\frac{\sin x^2}{x^2} = 1 - \frac{x^4}{3!} + \frac{x^8}{5!} - \frac{x^{12}}{7!} + \ldots$

17. **(B)**

The slope of the tangent line increases from $t = 5$ to $t = 7$ (since $N''(t) > 0$).
$N(5.5)$ is greater than $N(5) + N'(5)\Delta t = 22 + 12(.5) = 28$.
$N(5.5)$ is less than $N(5) + N'(7)\Delta t = 22 + 26(.5) = 35$.
So $28 < N(5.5) < 35$

18. **(B)**

$$\lim_{x \to 4} \frac{\int_{-4}^{x} f(t)\,dt}{f(x) + 8} = \frac{\int_{-4}^{4} f(t)\,dt}{f(4) + 8} = \frac{16 - 16}{-8 + 8} = \frac{0}{0}$$

L'Hospital's Rule: $\displaystyle \lim_{x \to 4} \frac{\int_{-4}^{x} f(t)\,dt}{f(x) + 8} = \lim_{x \to 4} \frac{\frac{d}{dx}\int_{-4}^{x} f(t)\,dt}{f'(x)} = \frac{f(4)}{f'(4)} = \frac{-8}{-2} = 4$

19. **(B)**

At first, the height goes up very quickly so $\frac{dh}{dt}$ is a very large number. But as time increases, the height doesn't go up as quickly as $\frac{dh}{dt}$ is smaller. $\frac{dh}{dt}$ decreases as the cone fills. Once the water reaches the height of the cylinder, the height increases at a steady rate, so $\frac{dh}{dt}$ is a constant.

The problem could be done analytically as well. For the cone, $V = \dfrac{\pi r^2 h}{3}$ with $r = \dfrac{h}{4}$ by similar triangles. $V = \dfrac{\pi h^3}{48}$ and $\dfrac{dV}{dt} = \dfrac{\pi h^2}{16}\dfrac{dh}{dt}$. Finally, $\dfrac{dh}{dt} = \dfrac{16}{\pi h^2}$. As h increases, $\dfrac{dh}{dt}$ decreases. The same type of analysis can be done for the cylinder. The distractor answer is (C), which represents a graph of the height as a function of time, not the change in height.

20. **(A)**

I. Ratio test: $\displaystyle\lim_{n\to\infty}\left|\dfrac{3^{n+1}}{n!}\cdot\dfrac{(n-1)!}{3^n}\right| = \lim_{n\to\infty}\left|\dfrac{3}{n}\right| = 0$ so $\displaystyle\sum_{n=1}^{\infty}\dfrac{3^n}{(n-1)!}$ is convergent.

II. Ratio test: $\displaystyle\lim_{n\to\infty}\left|\dfrac{(n+1)!}{(n+1)^3}\cdot\dfrac{n^3}{n!}\right| = \lim_{n\to\infty}\left|n+1\right| = \infty$ so $\displaystyle\sum_{n=1}^{\infty}\dfrac{n!}{n^3}$ is divergent.

III. Limit comparison test: $\displaystyle\lim_{n\to\infty}\left|\dfrac{\dfrac{n+3}{n^2+3n+2}}{\dfrac{1}{n}}\right| = \lim_{n\to\infty}\left|\dfrac{n^2+3n}{n^2+3n+2}\right| = 1$

Since the series $\dfrac{1}{n}$ is divergent, $\displaystyle\sum_{n=0}^{\infty}\dfrac{n+3}{(n+1)(n+2)}$ is divergent.

21. **(C)**

$\displaystyle\int_{-\infty}^{2} x^2 e^{x^3}\,dx$

$u = x^3 \qquad du = 3x^2\,dx$
$x = -\infty,\ u = -\infty \qquad x = 2,\ u = 8$

$\dfrac{1}{3}\displaystyle\int_{-\infty}^{2} 3x^2 e^{x^3}\,dx$

$\dfrac{1}{3}\displaystyle\int_{-\infty}^{8} e^u\,du$

$\dfrac{1}{3}\left[e^u\right]_{-\infty}^{8} = \dfrac{1}{3}\left(e^8 - \dfrac{1}{e^\infty}\right) = \dfrac{e^8}{3}$

22. **(C)**

$(-3,4): \dfrac{dy}{dx} = 9 - 12 + 2 = -1$ so $(-3,4)$ is not a critical point.

$(2,-3): \dfrac{dy}{dx} = 4 - 6 + 2 = 0$ so $(2,-3)$ is a critical point.

$\dfrac{d^2y}{dx^2} = 2x + x\dfrac{dy}{dx} + y$

$(2,-3): \dfrac{d^2y}{dx^2} = 4 + 2(0) - 3 = 1$ so $(2,-3)$ is a relative (local) minimum.

23. **(E)**

The form of a logistic DEQ is $\frac{dP}{dt} = kP(\text{carrying capacity} - P)$, eliminating choices A, B, and C.

Choice E: $2500 = \frac{1}{40000}(5000)(20000)$.

To do the problem directly, $\frac{dP}{dt} = kP(\text{carrying capacity} - P)$

$2500 = 5000k(25000 - 5000)$

$k = \frac{1}{2(20000)} = \frac{1}{40000}$

24. **(C)**

$P_3(x) \approx f(1) + f'(1)(x-1) + \frac{f''(1)}{2!}(x-1)^2 + \frac{f'''(1)}{3!}(x-1)^3$

$P_3\left(\frac{1}{2}\right) \approx 3 + 4\left(-\frac{1}{2}\right) - \frac{4}{2!}\left(-\frac{1}{2}\right)^2 - \frac{48}{3!}\left(-\frac{1}{2}\right)^3 = 3 - 2 - \frac{1}{2} + 1 = \frac{3}{2}$

25. **(D)**

The value of the stock is $10000 + \int_0^x S'(t)\, dt$.

There are 5 critical points ... values of t where $\frac{dS}{dt} = 0$.

$S(0) = 10000$

$10000 + \int_0^a S'(t)\, dt = 10000 + 4000 = 14000$

$10000 + \int_0^b S'(t)\, dt = 10000 + 4000 - 1200 = 12800$

$10000 + \int_0^c S'(t)\, dt = 10000 + 4000 - 1200 - 2500 = 10300$

$10000 + \int_0^d S'(t)\, dt = 10000 + 4000 - 1200 - 2500 + 2300 = 12600$

Maximum $= \$14,000$, Minimum $= \$10,000$, Difference $= \$4,000$.

Easier to find is the difference of the areas above the axis and the areas below the axis.

26. **(D)**

Integration by parts:

$$\int \frac{x^2}{1+x^2}\,dx \qquad \begin{aligned} u &= x^2 \qquad\qquad v = \tan^{-1} x \\ du &= 2x\,dx \quad dv = \frac{1}{1+x^2}\,dx \end{aligned}$$

$$\int \frac{x^2}{1+x^2}\,dx = x^2 \tan^{-1} x - 2\int x \tan^{-1} x\,dx$$

So $f(x) = \tan^{-1} x$

27. **(B)**

$$x = r\cos\theta = e^{\theta}\cos\theta \qquad\qquad y = r\sin\theta = e^{\theta}\sin\theta$$

$$\frac{dx}{d\theta} = e^{\theta}(\cos\theta - \sin\theta) \qquad \frac{dy}{d\theta} = e^{\theta}(\sin\theta + \cos\theta)$$

$$\frac{dy}{dx} = \frac{e^{\theta}(\sin\theta + \cos\theta)}{e^{\theta}(\cos\theta - \sin\theta)} \Rightarrow \frac{dy}{dx}_{\theta = \frac{\pi}{2}} = \frac{1}{-1} = -1$$

$$x = e^{\pi/2}\cos\frac{\pi}{2} = 0 \qquad\qquad y = e^{\pi/2}\sin\frac{\pi}{2} = e^{\pi/2}$$

$$y - e^{\pi/2} = -1(x - 0) \Rightarrow y = e^{\pi/2} - x$$

28. **(A)**

$$y = 15 - 15x \Rightarrow 15x = 15 - y \Rightarrow x = 1 - \frac{y}{15}$$

Base: $V = \pi \int_{0}^{10} \left(1 - \frac{y}{15}\right)^2 dy$

Light: $15 - 15x = 10 \Rightarrow x = \frac{1}{3}$, which is the radius.

Volume of light $= \pi r^2 h = \pi \left(\frac{1}{3}\right)^2 (2) = \frac{2\pi}{9}$

Total Volume: $V = \pi \int_{0}^{10} \left(1 - \frac{y}{15}\right)^2 dy + \frac{2\pi}{9}$

Detailed Explanations of Answers

Section I, Part B

76. **(B)**

A. $f(a)$ is positive

B. $f'(a)$ is the slope of the tangent line to f at a and is negative

C. $\lim\limits_{x\to b}\dfrac{f(x)-f(b)}{x-b}$ is asking for $f'(b)$. This is negative, but the slope at a is steeper and thus a smaller number.

D. $\dfrac{f(b)-f(a)}{b-a}$ is the slope of the line between a and b. This is negative, but $f'(a)$ is a smaller number.

E. $f''(b)$ is the concavity at b. Since the curve is concave up, this is positive.

77. **(D)**

$$f(x)=0+0-\frac{3(x+1)^2}{2!}+\frac{8(x+1)^3}{3!}+\dots$$

$$\lim_{x\to-1}\frac{f(x)}{\cos(\pi x)+1}=\frac{0}{\cos(-\pi)+1}=\frac{0}{0}$$

$$f'(x)=-3(x+1)+4(x+1)^2+\dots$$

L'Hospital's Rule: $\lim\limits_{x\to-1}\dfrac{f'(x)}{-\pi\sin(\pi x)}=\dfrac{0}{-\pi\sin(-\pi)}=\dfrac{0}{0}$

$$f''(x)=-3+8(x+1)+\dots$$

L'Hospital's Rule: $\lim\limits_{x\to-1}\dfrac{f''(x)}{-\pi^2\cos(\pi x)}=\dfrac{-3}{-\pi^2\cos(-\pi)}=\dfrac{-3}{\pi^2}$

78. **(A)**

$$r'(t) = 100\left(\frac{1}{2\sqrt{t}} + 2\cos t\right) = 0$$

The graph of $r'(t) = 0$ gives $t = 1.760,\ 4.596$

Since the change is negative, we want the value of t when $r(t)$ is a minimum.

$r(0) = -400 \quad r(1.760) \approx -71 \quad r(4.596) \approx -384 \quad r(6) \approx -210$

At $t = 0$, the stack is decreasing by 400 coupons an hour.

79. **(D)**

I is true because $\lim_{x \to 0} f(x) = 2.01 - \sqrt{0.0001} = 2.01 - 0.01 = 2$

II is true because $f(0) = 2 = \lim_{x \to 0} f(x)$.

III Looks are deceiving. It appears that there is a cusp point at $x = 0$ and thus not differentiable.

But $f'(x) = \dfrac{2\sin x \cos x}{-2\sqrt{\sin^2 x + 0.0001}}$ and $f'(0) = 0$ so f is differentiable at $x = 0$.

80. **(B)**

This problem has what are called "distractors" in it. The table is meant to distract you from finding the value of the definite integral $\int_0^1 (e^x - e^{-x})\, dx$.

$$\int_0^1 (e^x - e^{-x})\, dx = \left[e^x + e^{-x}\right]_0^1$$

$$= e + \frac{1}{e} - (1+1) = e + \frac{1}{e} - 2 = 2.718 + \frac{1}{2.718} - 2 = 0.718 + \frac{1}{2.718} \approx 1.086$$

Since this problem is on the calculator section, the calculator can be used to find $\int_0^1 (e^x - e^{-x})\, dx$.

81. **(C)**

$$\sin x = x - \frac{x^3}{3!} + \frac{x^5}{5!} - \frac{x^7}{7!} \ldots \Rightarrow x\sin x = x^2 - \frac{x^4}{3!} + \frac{x^6}{5!} - \frac{x^8}{7!} + \ldots$$

$$\int_0^1 \left(x^2 - \frac{x^4}{3!} + \frac{x^6}{5!} - \frac{x^8}{7!} + \ldots\right) dx = \left[\frac{x^3}{3} - \frac{x^5}{5\cdot 3!} + \frac{x^7}{7\cdot 5!} - \frac{x^9}{9\cdot 7!} + \ldots\right]_0^1$$

$$\int_0^1 x\sin x\, dx = \frac{1}{3} - \frac{1}{30} + \frac{1}{840} - \frac{1}{45360} - \ldots$$

Since this is an alternating series and $\dfrac{1}{45360} < 0.001$, 3 terms are needed.

82. **(B)**

$$f'(x) = \cos x(-\cos x) + (1 - \sin x)(-\sin x) = \sin^2 x - \cos^2 x - \sin x$$

$$\sin^2 x - \cos^2 x - \sin x = \frac{f(\pi) - f(0)}{\pi - 0} = \frac{0 - 2}{\pi} = \frac{-2}{\pi}$$

By calculator: $x = 0.839, 2.302$

$$\int_{0.839}^{2.302} f(x)\,dx = 1.463$$

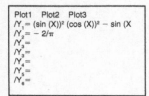 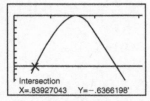 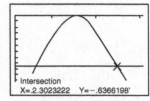 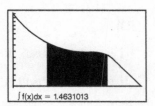

83. **(A)**

$$\frac{f'(0)}{1!} = \frac{1}{2!} \Rightarrow f'(0) = \frac{1!}{2!}$$

$$\frac{f'''(0)}{3!} = -\frac{1}{6!} \Rightarrow f'''(0) = -\frac{3!}{6!}$$

$$\frac{f^{(5)}(0)}{5!} = \frac{1}{10!} \Rightarrow f'''(0) = \frac{5!}{10!}$$

$$\vdots$$

$$\frac{f^{(25)}(0)}{25!} = \frac{1}{50!} \Rightarrow f^{(25)}(0) = \frac{25!}{50!}$$

84. **(E)**

The p-series $\displaystyle\sum_{n=1}^{\infty} \frac{n}{n^{3p-4}} = \sum_{n=1}^{\infty} \frac{1}{n^{3p-5}}$ will be convergent when $3p - 5 > 1$ or $p > 2$.

When $p = 2$, the series is an alternating harmonic series which converges so $p \geq 2$.

Since $\displaystyle\sum_{n=1}^{\infty} \frac{n}{n^{3p-4} + 1}$ will contain fractions less than $\displaystyle\sum_{n=1}^{\infty} \frac{n}{n^{3p-4}}$, convergence also occurs

when $p \geq 2$.

85. **(E)**

The side of the square is $2r$ so the radius of the circumscribed circle is $r\sqrt{2}$.

$A_{\text{inscribed}} = \pi r^2$ $C_{\text{circumscribed}} = 2\pi r\sqrt{2}$

$\dfrac{dA}{dt} = 2\pi r \dfrac{dr}{dt}$ $\dfrac{dC}{dt} = 2\pi\sqrt{2}\,\dfrac{dr}{dt}$

$2\pi r \dfrac{dr}{dt} = 2\pi\sqrt{2}\,\dfrac{dr}{dt}$

$r = \sqrt{2}$

$A_{\text{square}} = \left(2\sqrt{2}\right)^2 = 8$

86. **(D)**

$\lim\limits_{n\to\infty}\left|\dfrac{x^{n+2}}{(n+1)(n+2)}\cdot\dfrac{n(n+1)}{x^{n+1}}\right| = |x| < 1$ so $-1 < x < 1$

$x = -1: \dfrac{1}{2} - \dfrac{1}{6} + \dfrac{1}{12} + \dots$ Convergent alternating series

$x = 1: \dfrac{1}{2} + \dfrac{1}{6} + \dfrac{1}{12} + \dots$ Convergent $-$ compared to $\dfrac{1}{n^2}$

Note that $\int f(x)\,dx$ will have the same interval of convergence as the summation.

87. **(B)**

$L = e^{-x/2} - \dfrac{x}{2}.$

In an isosceles right triangle, the legs are the same so the area $= \dfrac{1}{2}L^2.$

Thus, $V = \dfrac{1}{2}\displaystyle\int_0^{1.134}\left(e^{-x/2} - \dfrac{x}{2}\right)^2 dx = 0.178$

88. **(C)**

The maximum value of $f^{(4)}(x) = e^{\sin x + \cos x}$ is 4.113 as shown by the graph below. (Note that although using the calculator's MAXIMUM function is not permitted on a justification problem, that is not an issue with multiple-choice questions.) So the Lagrange error is $\dfrac{4.113}{4!} = 0.171.$

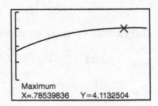

Maximum
X=.78539836   Y=4.1132504

89. **(B)**

Quadrant I: $A = .5 \int_{0}^{\pi/2} (2\theta + 3\cos\theta)^2 d\theta = 9.543$

Quadrant II: $A = .5 \int_{\pi/2}^{\pi} (2\theta + 3\cos\theta)^2 d\theta = 6.197$

Difference: 3.346

90. **(C)**

Directional changes occur when velocity switches from positive to negative or from negative to positive. This occurs at time $t = 1$ and 4 only.

91. **(B)**

This is an integration by parts problem.

$u = x \qquad v = f(x)$

$du = dx \quad dv = f'(x)dx$

$\int_{-1}^{6} x \cdot f'(x) \, dx = [x \cdot f(x)]_{-1}^{6} - \int_{-1}^{6} f(x) \, dx$

$= 6(-4) - (-1)(3) - 15$

$= -36$

92. **(D)**

$v(t) = \left\langle \int 2\pi \sin\pi t \ dt, \ \int e^t \, dt \right\rangle = \left\langle -2\cos\pi t + C_1, e^t + C_2 \right\rangle$

$v(0) = \left\langle -2(1) + C_1, 1 + C_2 \right\rangle = \left\langle 0,0 \right\rangle \quad C_1 = 2, C_2 = -1$

$v(t) = \left\langle -2\cos\pi t + 2, \ e^t - 1 \right\rangle$

$\text{Speed}_{t=1} = \sqrt{(4)^2 + (e-1)^2} = \sqrt{18.952} = 4.353$

Detailed Explanations of Answers

Section II, Part A

Question 1

(a) $E'(t) = -50t \Rightarrow E'(1.5) = -50(1.5) = -75$.

The rate of change of steaks being eaten is decreasing at 75 steaks per hour per hour, or $75 \dfrac{\text{steaks}}{\text{hr}^2}$.

(b) $\dfrac{\int\limits_0^4 E(t)\,dt}{4} = \dfrac{\int\limits_0^4 \left(400 - 25t^2\right)dt}{4} \approx 267 \dfrac{\text{steaks}}{\text{hour}}$.

(c) Since $C(t)$ and $E(t)$ are rates, they are derivatives. We are interested in where $C(t) - E(t) = 0$ (the rate of steaks being grilled equals the rate of steaks eaten).
$400 - 25t^2 = 364 \Rightarrow t = 1.2$ $\qquad$ $400 - 25t^2 = 144 \Rightarrow t = 3.2$

So the candidates for the absolute minimum are at $t = 0$, $t = 1.2$, $t = 3.2$, $t = 4$

t (hours)	Steaks Grilling or Warming
0	150
1.2	$150 + 364(1.2) - \int\limits_0^{1.2} (400 - 25t^2)\,dt \approx 121$
3.2	$150 + 364(2) + 144(1.2) - \int\limits_0^{3.2} (400 - 25t^2)\,dt \approx 44$
4	$150 + 364(2) + 144(2) - \int\limits_0^{4} (400 - 25t^2)\,dt \approx 99$

At $t = 3.2$ hours, there will be a minimum number of steaks that either have been grilled or are in the warmers, but not eaten.

(d) At 4 hours, there are 99 steaks remaining.

Steaks eaten between hour 4 and hour k: $\displaystyle\int_{4}^{4+k} \left(50\sqrt{t}-4\right)dt = 99$

Question 2

(a) $\displaystyle\frac{dx}{dt}\bigg|_{t=5} = 5-\cos^2 5 = 4.920$

The particle is moving to the right at $t = 5$ because $\displaystyle\frac{dx}{dt}\bigg|_{t=5} > 0.$

$\displaystyle\frac{dy}{dt}\bigg|_{t=5} = \frac{1-4\ln 6}{6} = -1.028$

$\displaystyle\frac{dy}{dx} = \frac{dy/dt\big|_{t=5}}{dx/dt\big|_{t=5}} = \frac{-1.028}{4.920} = -0.209$

(b) $\displaystyle x(5) = -\pi + \int_{1}^{5}\left(t-\cos^2 t\right)dt = -\pi + 10.363 = 7.222$

$\displaystyle y(5) = -1 + \int_{1}^{5}\left(\frac{1-4\ln(t+1)}{t+1}\right)dt = -1-4.361 = -5.361$

Position: $(7.222, -5.361)$

(c) Speed $= \sqrt{\left[x'(5)\right]^2 + \left[y'(5)\right]^2} = \sqrt{4.920^2 + (-1.028)^2} = 5.026$

$$a(t) = \left\langle 1+2\sin t\cos t, \frac{(t+1)\left(\dfrac{-4}{t+1}\right)-\left[1-4\ln(t+1)\right]}{(t+1)^2}\right\rangle$$

$a(5) = \langle 0.456, 0.060\rangle$

Note that it is permissible to use the calculator's nDeriv function to compute these answers rather than take the derivatives.

(d) Distance $= \displaystyle\int_{0}^{5}\sqrt{\left[x'(t)\right]^2 + \left[y'(t)\right]^2}\,dt = 12.161$

Detailed Explanations of Answers

Section II, Part B

Question 3

(a) $y = \dfrac{2x}{3} \implies x = \dfrac{3y}{2}$

$\dfrac{3y}{2} = \dfrac{\sqrt{y^2 + 32}}{2} \implies 3y = \sqrt{y^2 + 32}$

$9y^2 = y^2 + 32 \implies y = 2,\ x = 3$, so point P is $(3, 2)$.

$\dfrac{dx}{dy} = \dfrac{1}{2}\left(\dfrac{2y}{2\sqrt{y^2 + 32}} \right) = \dfrac{y}{2\sqrt{y^2 + 32}} \implies \dfrac{dx}{dy}\Big|_{y=2} = \dfrac{2}{12} = \dfrac{1}{6}$

(b) Radius $= \dfrac{1}{2}\left(\dfrac{\sqrt{y^2 + 32}}{2} - \dfrac{3y}{2} \right) = \dfrac{1}{4}\left(\sqrt{y^2 + 32} - 3y \right)$

$\text{Volume} = \dfrac{\pi}{2}\int_0^2 \left[\dfrac{1}{4}\left(\sqrt{y^2 + 32} - 3y \right) \right]^2 dy$ OR

$\text{Volume} = \dfrac{\pi}{32}\int_0^2 \left(\sqrt{y^2 + 32} - 3y \right)^2 dy$

(c) $x = \dfrac{\sqrt{y^2 + 32}}{2} \implies 2x = \sqrt{y^2 + 32} \implies 4x^2 - y^2 = 32$

Since $x = r\cos\theta$ and $y = r\sin\theta$,

$4r^2\cos^2\theta - r^2\sin^2\theta = 32$

$r^2 = \dfrac{32}{4\cos^2\theta - \sin^2\theta}$

(d) If α represents the angle that the line makes with the x-axis, then $\tan\alpha = \dfrac{2}{3}$.

$\text{Area} = \dfrac{1}{2}\int_0^{\tan^{-1}(2/3)} r^2\, d\theta = \dfrac{1}{2}\int_0^{\tan^{-1}(2/3)} \dfrac{32}{4\cos^2\theta - \sin^2\theta}\, d\theta$ OR $\int_0^{\tan^{-1}(2/3)} \dfrac{16}{4\cos^2\theta - \sin^2\theta}\, d\theta$

Question 4

(a)

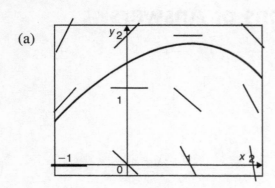

(b) $\dfrac{dy}{dx} = 0 \Rightarrow y - x - 1 = 0$

$y - e - 1 = 0 \Rightarrow y = e + 1$

(c)

x	1	0.75	0.5
$y_{new} = y_{old} + \dfrac{dy}{dx}\Delta x$	2	$2 + (0)(-0.25) = 2$	$2 + (0.25)(-0.25) = \boxed{\dfrac{31}{16}}$
$\dfrac{dy}{dx}$	0	$2 - 0.75 - 1 = 0.25$	

(d) $\dfrac{dy}{dx} = y - x - 1 > 0 \Rightarrow y > x + 1$

$\dfrac{d^2y}{dx^2} = \dfrac{dy}{dx} - 1 = y - x - 1 - 1 < 0 = y - x - 2 < 0 \Rightarrow y < x + 2$

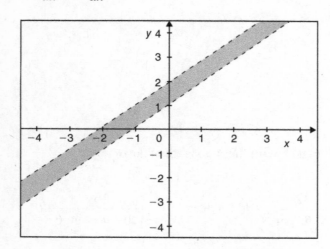

$$\frac{d^2y}{dx^2}\bigg|_{(1,2)} = y - x - 2 = 2 - 1 - 2 = -1$$

(e) $P_2(x) = 2 + 0(x-1) - \frac{1(x-1)^2}{2!} = 2 - \frac{(x-1)^2}{2}.$

Question 5

(a) $v'(t) = -t^2 e^{-t} + 2te^{-t}$

$$v'(t) = \frac{t(2-t)}{e^t} = 0$$

$t = 2$ because $v'(t)$ switches from positive to negative at that value.

(b) $v(t) = \frac{t^2}{e^t}$ which is always positive.

$$a(t) = \frac{2t(2-t)}{e^t}$$ which is always negative if $t > 2$.

Since as t gets larger, $v(t) > 0$ and $a(t) < 0$, the particle is slowing down.

(c) $x(t) = \int v(t)\, dt = \int t^2 e^{-t}\, dt$ $u = t^2$ $v = -e^{-t}$

$x(t) = -t^2 e^{-t} + 2\int te^{-t}\, dt$ $du = 2t\, dt$ $dv = e^{-t} dt$

                               $u = t$ $v = -e^{-t}$

$x(t) = -t^2 e^{-t} + 2\left(-te^{-t} - \int -e^{-t}\, dt\right)$ $du = dt$ $dv = e^{-t} dt$

$x(t) = -t^2 e^{-t} - 2te^{-t} - 2e^{-t} + C$

$x(t) = \frac{-t^2 - 2t - 2}{e^t} + C$

$x(0) = -2 + C = 1 \Rightarrow C = 3$

$x(t) = \frac{-t^2 - 2t - 2}{e^t} + 3$

$x(2) = \frac{-10}{e^2} + 3$

(d) Since $v(t) > 0$, distance $= \displaystyle\int_2^\infty v(t)\, dt = \left[\frac{-t^2 - 2t - 2}{e^t}\right]_2^\infty$

distance $= 0 \ (\text{by L'Hospital's Rule}) - \left(\frac{-10}{e^2}\right) = \frac{10}{e^2}$

Question 6

(a) This is a geometric series with ratio $r = -\left(x + \dfrac{1}{2}\right)$

This converges when $\left|x + \dfrac{1}{2}\right| < 1$ so $\dfrac{-3}{2} < x < \dfrac{1}{2}$.

OR $\displaystyle\lim_{n\to\infty} \left|\dfrac{(-1)^{n+1}\left(x + \dfrac{1}{2}\right)^{n+1}}{(-1)^{n}\left(x + \dfrac{1}{2}\right)^{n}}\right| < 1 = \left|x + \dfrac{1}{2}\right| < 1 \Rightarrow \dfrac{-3}{2} < x < \dfrac{1}{2}.$

$x = \dfrac{-3}{2} : \displaystyle\sum_{n=0}^{\infty}(-1)^{n}\left(x + \dfrac{1}{2}\right)^{n} = 1 + 1 + 1 + \ldots \Rightarrow$ divergent

$x = \dfrac{1}{2} : \displaystyle\sum_{n=0}^{\infty}(-1)^{n}\left(x + \dfrac{1}{2}\right)^{n} = 1 - 1 + 1 + \ldots \Rightarrow$ divergent

The interval of convergence is $\dfrac{-3}{2} < x < \dfrac{1}{2}$.

(b) Since the series is geometric with $r = -\left(x + \dfrac{1}{2}\right)$

$\displaystyle\sum_{n=0}^{\infty}(-1)^{n}\left(x + \dfrac{1}{2}\right)^{n} = \dfrac{1}{1 - \left[-\left(x + \dfrac{1}{2}\right)\right]}$

$= \dfrac{1}{1 + x + \dfrac{1}{2}} = \dfrac{2}{2 + 2x + 1} = \dfrac{2}{2x + 3}$ for $-\dfrac{3}{2} < x < \dfrac{1}{2}$

(c) $g(x) = f\left(x^2 - \dfrac{1}{2}\right) = 1 - x^2 + x^4 \ldots + (-1)^{n} x^{2n} + \ldots$

$g\left(\sqrt{\dfrac{1}{3}}\right) = 1 - \left(\dfrac{1}{3}\right) + \left(\dfrac{1}{3}\right)^{2} - \left(\dfrac{1}{3}\right)^{3} + \ldots = \dfrac{1}{1 + \dfrac{1}{3}} = \dfrac{3}{3 + 1} = \dfrac{3}{4}.$

AP Calculus AB
Answer Sheet

Section I, Part A

1. Ⓐ Ⓑ Ⓒ Ⓓ Ⓔ
2. Ⓐ Ⓑ Ⓒ Ⓓ Ⓔ
3. Ⓐ Ⓑ Ⓒ Ⓓ Ⓔ
4. Ⓐ Ⓑ Ⓒ Ⓓ Ⓔ
5. Ⓐ Ⓑ Ⓒ Ⓓ Ⓔ
6. Ⓐ Ⓑ Ⓒ Ⓓ Ⓔ
7. Ⓐ Ⓑ Ⓒ Ⓓ Ⓔ
8. Ⓐ Ⓑ Ⓒ Ⓓ Ⓔ
9. Ⓐ Ⓑ Ⓒ Ⓓ Ⓔ
10. Ⓐ Ⓑ Ⓒ Ⓓ Ⓔ
11. Ⓐ Ⓑ Ⓒ Ⓓ Ⓔ
12. Ⓐ Ⓑ Ⓒ Ⓓ Ⓔ
13. Ⓐ Ⓑ Ⓒ Ⓓ Ⓔ
14. Ⓐ Ⓑ Ⓒ Ⓓ Ⓔ

15. Ⓐ Ⓑ Ⓒ Ⓓ Ⓔ
16. Ⓐ Ⓑ Ⓒ Ⓓ Ⓔ
17. Ⓐ Ⓑ Ⓒ Ⓓ Ⓔ
18. Ⓐ Ⓑ Ⓒ Ⓓ Ⓔ
19. Ⓐ Ⓑ Ⓒ Ⓓ Ⓔ
20. Ⓐ Ⓑ Ⓒ Ⓓ Ⓔ
21. Ⓐ Ⓑ Ⓒ Ⓓ Ⓔ
22. Ⓐ Ⓑ Ⓒ Ⓓ Ⓔ
23. Ⓐ Ⓑ Ⓒ Ⓓ Ⓔ
24. Ⓐ Ⓑ Ⓒ Ⓓ Ⓔ
25. Ⓐ Ⓑ Ⓒ Ⓓ Ⓔ
26. Ⓐ Ⓑ Ⓒ Ⓓ Ⓔ
27. Ⓐ Ⓑ Ⓒ Ⓓ Ⓔ
28. Ⓐ Ⓑ Ⓒ Ⓓ Ⓔ

Section I, Part B

76. Ⓐ Ⓑ Ⓒ Ⓓ Ⓔ
77. Ⓐ Ⓑ Ⓒ Ⓓ Ⓔ
78. Ⓐ Ⓑ Ⓒ Ⓓ Ⓔ
79. Ⓐ Ⓑ Ⓒ Ⓓ Ⓔ
80. Ⓐ Ⓑ Ⓒ Ⓓ Ⓔ
81. Ⓐ Ⓑ Ⓒ Ⓓ Ⓔ
82. Ⓐ Ⓑ Ⓒ Ⓓ Ⓔ
83. Ⓐ Ⓑ Ⓒ Ⓓ Ⓔ
84. Ⓐ Ⓑ Ⓒ Ⓓ Ⓔ
85. Ⓐ Ⓑ Ⓒ Ⓓ Ⓔ
86. Ⓐ Ⓑ Ⓒ Ⓓ Ⓔ
87. Ⓐ Ⓑ Ⓒ Ⓓ Ⓔ
88. Ⓐ Ⓑ Ⓒ Ⓓ Ⓔ
89. Ⓐ Ⓑ Ⓒ Ⓓ Ⓔ
90. Ⓐ Ⓑ Ⓒ Ⓓ Ⓔ
91. Ⓐ Ⓑ Ⓒ Ⓓ Ⓔ
92. Ⓐ Ⓑ Ⓒ Ⓓ Ⓔ

FREE-RESPONSE ANSWER SHEET

For the free-response section, write your answers on sheets of blank paper.

AP Calculus BC
Answer Sheet

Section I, Part A

1. Ⓐ Ⓑ Ⓒ Ⓓ Ⓔ
2. Ⓐ Ⓑ Ⓒ Ⓓ Ⓔ
3. Ⓐ Ⓑ Ⓒ Ⓓ Ⓔ
4. Ⓐ Ⓑ Ⓒ Ⓓ Ⓔ
5. Ⓐ Ⓑ Ⓒ Ⓓ Ⓔ
6. Ⓐ Ⓑ Ⓒ Ⓓ Ⓔ
7. Ⓐ Ⓑ Ⓒ Ⓓ Ⓔ
8. Ⓐ Ⓑ Ⓒ Ⓓ Ⓔ
9. Ⓐ Ⓑ Ⓒ Ⓓ Ⓔ
10. Ⓐ Ⓑ Ⓒ Ⓓ Ⓔ
11. Ⓐ Ⓑ Ⓒ Ⓓ Ⓔ
12. Ⓐ Ⓑ Ⓒ Ⓓ Ⓔ
13. Ⓐ Ⓑ Ⓒ Ⓓ Ⓔ
14. Ⓐ Ⓑ Ⓒ Ⓓ Ⓔ

15. Ⓐ Ⓑ Ⓒ Ⓓ Ⓔ
16. Ⓐ Ⓑ Ⓒ Ⓓ Ⓔ
17. Ⓐ Ⓑ Ⓒ Ⓓ Ⓔ
18. Ⓐ Ⓑ Ⓒ Ⓓ Ⓔ
19. Ⓐ Ⓑ Ⓒ Ⓓ Ⓔ
20. Ⓐ Ⓑ Ⓒ Ⓓ Ⓔ
21. Ⓐ Ⓑ Ⓒ Ⓓ Ⓔ
22. Ⓐ Ⓑ Ⓒ Ⓓ Ⓔ
23. Ⓐ Ⓑ Ⓒ Ⓓ Ⓔ
24. Ⓐ Ⓑ Ⓒ Ⓓ Ⓔ
25. Ⓐ Ⓑ Ⓒ Ⓓ Ⓔ
26. Ⓐ Ⓑ Ⓒ Ⓓ Ⓔ
27. Ⓐ Ⓑ Ⓒ Ⓓ Ⓔ
28. Ⓐ Ⓑ Ⓒ Ⓓ Ⓔ

Section I, Part B

76. Ⓐ Ⓑ Ⓒ Ⓓ Ⓔ
77. Ⓐ Ⓑ Ⓒ Ⓓ Ⓔ
78. Ⓐ Ⓑ Ⓒ Ⓓ Ⓔ
79. Ⓐ Ⓑ Ⓒ Ⓓ Ⓔ
80. Ⓐ Ⓑ Ⓒ Ⓓ Ⓔ
81. Ⓐ Ⓑ Ⓒ Ⓓ Ⓔ
82. Ⓐ Ⓑ Ⓒ Ⓓ Ⓔ
83. Ⓐ Ⓑ Ⓒ Ⓓ Ⓔ
84. Ⓐ Ⓑ Ⓒ Ⓓ Ⓔ
85. Ⓐ Ⓑ Ⓒ Ⓓ Ⓔ
86. Ⓐ Ⓑ Ⓒ Ⓓ Ⓔ
87. Ⓐ Ⓑ Ⓒ Ⓓ Ⓔ
88. Ⓐ Ⓑ Ⓒ Ⓓ Ⓔ
89. Ⓐ Ⓑ Ⓒ Ⓓ Ⓔ
90. Ⓐ Ⓑ Ⓒ Ⓓ Ⓔ
91. Ⓐ Ⓑ Ⓒ Ⓓ Ⓔ
92. Ⓐ Ⓑ Ⓒ Ⓓ Ⓔ

FREE-RESPONSE ANSWER SHEET

For the free-response section, write your answers on sheets of blank paper.

Index

NOTES

NOTES